Immunoassay Technology
Volume 2

Immunoassay Technology

Volume 2

Editor S. B. Pal

W DE G

Walter de Gruyter · Berlin · New York 1986

Editor

S. B. Pal, D. Phil., Dr. rer. biol. hum. M. I. Biol.
Universität Ulm
Department für Innere Medizin
Steinhövelstraße 9
D-7900 Ulm
F. R. of Germany

ISBN 3 11 010948 4 Walter de Gruyter · Berlin · New York
ISBN 0-89925-195-1 Walter de Gruyter, Inc., New York

 Printing: Gerike GmbH, Berlin. – Binding: Dieter Mikolai, Berlin. – Cover-Design: Hans Bernd Lindemann, Berlin – Printed in Germany.

Preface

The second volume of this periodical, Immunoassay Technology, contains several articles which it is hoped readers will find as interesting, useful and thought-provoking as those presented in the previous volume, in particular to recent entrants in this field. I take this opportunity to thank Dr. D. Donaldson, M. R. C. P., F. R. C. Path., Consultant Chemical Pathologist, East Surrey Hospital, Redhill, Surrey, U. K., for his helpful suggestions during the editing of this volume.
Much credit goes to Walter de Gruyter Publishers - Science Division - for making the necessary arrangements for the rapid publication of this volume. I also thank Mrs. M. R. Pal for her assistance as an honorary editorial secretary.

December 1985 S. B. Pal

In Memoriam

Professor Michael Finkelstein
15. 12. 1916 – 27. 8. 1985

During the gestation period of Immunoassay Technology, many colleagues provided me with various ideas and suggestions. I very warmly remember the helpful encouragement given to me by Michael Finkelstein, who suddenly passed away on 27th August 1985, and to whose memory this volume is dedicated.
Michael Finkelstein was born in Sosnowiec, Poland, and after completing his High School education there, he emigrated to Palestine in 1934, and enrolled at the Hebrew University for the study of Biology (Major in Biochemistry), obtaining an MSc Degree, later culminating in a Ph. D. in Endocrinology from the same University.
His scientific career started as a Research Assistant in Paediatrics (1940) at the Hadassah Hospital and as a Demonstrator in Biochemistry at the Hebrew University, Jerusalem, where he eventually became Professor and Chairman, Department of Endocrinology. During his working period at the Hebrew University, he went abroad with short-term appointments to the Universities of Wales, Basel, Sidney, and to the Worcester Foundation, Shrewsbury, Massachusetts, U.S.A. He was also a Consultant in Human Reproduction, World Health Organisation, posted in New Delhi, India, for a year, and subsequently went as a visiting Professor in Biochemistry to the University of Louisville in Kentucky, U.S.A.
Michael Finkelstein received many awards from various countries, including his own. He was one of the founders and Executive Member of the International Study

Group for Steroid Hormones, Rome, a Corresponding editor for the International Journal "Steroids", Co-editor, "Research on Steroids", volumes 3, 4 and 5, and he published 95 scientific papers and reviews, including a review article in volume 3 of "Hormones in Normal and Abnormal Human Tissues" edited by K. Fotherby and S. B. Pal.

Michael Finkelstein was happily married and had a son and a daughter, who are now well placed in life.

Members of the World Scientific Community who knew him, found in him a good scientist and a very kind and warm human being who will be sadly missed. The editor of this periodical had the pleasure of knowing him for many years and feels a real sense of loss at his untimely death.

Contents

CONJUGATION PROCEDURES IN ENZYME IMMUNOASSAY

D. G. Williams
Department of Clinical Biochemistry, District General Hospital,
Kayll Road, Sunderland, U.K.

Summary

One of the requirements for enzyme immunoassay (EIA) is the labelling of one of the components of the assay (the ligand, the binder, or the second binder) with an enzyme. This must be done in such a manner that the immunological activity of the binder and the catalytic activity of the enzyme are retained. There are many methods by which this can be carried out, but unlike radioimmunoassay, there is no method which is universally applicable. This article will consider some of the methods that have been used, and will also indicate techniques which may be of use to those who wish to form their own conjugates.

Introduction

The labelling of a component of the assay system with an enzyme is an essential requirement for a functional immunoassay. There are several types of assay in which different components of the system are labelled. The following list is a brief summary of the components required:

(a) EMIT (Syva Corporation, Palo Alto, California, U.S.A.) (1)

This is a "homogeneous" system in which the ligand is labelled. This system is for haptens (i.e. compounds with a molecular

Immunoassay Technology, Vol. 2

weight of less than 1000 daltons) and requires special conjugation procedures.

(b) ELISA (enzyme linked immunosorbent assay)

There are several variants of this type of assay, the common feature being the immobilization of one component of the assay system to a solid phase support.

(i) Assay for circulating antibodies (2). The labelled component is a host specific antiglobulin directed against one species, e.g. goat anti-human globulin.

(ii) Assay for a ligand by double antibody (sandwich method)(3). In this method, the antibody specific for the ligand must be labelled. The ligand must have two binding sites.

(iii) Assay using labelled ligand (4). The ligand is labelled and competes with free ligand for the immobilized binder.

(c) EIA (enzyme immunoassay) (5)

There are many examples of this type of assay which require labelled ligand.

It can be seen, therefore, that several components of the assay system could be conjugated, depending on the type of assay required.

Conjugation requires the formation of a stable covalent bond between the enzyme and the other molecule. This is achieved by use of a reagent that will react with functional groups on both molecules. These groups include amide, amino, phenyl, carboxyl, hydroxyl and thiol which are commonly found in proteins. There exists a wide variety of reagents that can be used to link protein to protein. Haptens usually require modification to introduce a reactive group, which can then be further reacted to form a conjugate with the enzyme. Similar techniques are used to couple haptens to carrier proteins for immunization purposes, and a large body of literature exists on this topic. Although this review will concentrate mainly on linking enzymes to large molecules, some attention will be given to the formation of hapten-enzyme conjugates.

Certain labelled compounds are now commercially available. EMIT (TM) methods come ready to use in kit form, and labelled second antibodies directed against immunoglobulins from a certain species are also available. Thus alkaline phosphatase or peroxidase labelled antibodies against human, canine, murine, goat and sheep immunoglobulins can be purchased (Miles Research Products, Stoke Poges, U.K.). There still remain many assays where the binder or the ligand has to be labelled, and the ways in which this can be achieved will now be considered.

Types of Conjugation Reagents

In a comprehensive review of protein-protein conjugation reactions (6), a wide variety of reagents which can form conjugates are listed. Four classes of reaction are considered (Table 1).

Table 1. Types of conjugation reaction

	Type	Example of reagent
1	One-stage reaction	Glutaraldehyde
2	Two-stage reaction	Toluene 2,4 diisocyanate (TDIC)
3	Three-stage reaction	bis diazotised p-phenylenediamine
4	Protected reaction	Glutaraldehyde plus protecting agent

Of the above, only the one- and two-stage reactions have been widely used to form conjugates. Type 3, as its name implies, involves reacting the two separate proteins (ligand and enzyme) with two separate reagents, before allowing the substituted proteins to react to form a stable link. This does not appear to have been used to form enzyme labelled compounds.

In Type 4 reactions, the immunological or catalytic

properties of one of the proteins is protected during the reaction. Thus, an immunosorbent was used to protect the antigen combining site of Fab fragments during conjugation to peroxidase which had been treated with glutaraldehyde (7). This method does not appear to be widely used. We shall now consider one- and two-stage reactions.

One-Stage Reactions

These reactions depend on the addition of the reagent to a mixture of the two proteins to be conjugated. The reagent most commonly used is glutaraldehyde [$OHC.(CH_2)_4.CHO$] which links free amino groups on both proteins. It was first applied (8), using the conjugates for the detection of cellular antigens and antibodies. However, it was not until later (9, 10) that these conjugates were used in assays.

Another group of reagents are the carbodiimides (general formula $R.N = C = N.R^1$ where R and R^1 are alkyl or aryl groups) which have been used to link bradykinin and angiotensin to albumin (11). Attempts to use the reagent to link peroxidase and alkaline phosphatase to human placental lactogen were unsuccessful owing to a considerable loss of enzyme activity (12).

Although easy to perform, the method suffers from a major drawback. As all three reagents are present in the reaction mixture, protein-protein and enzyme-enzyme conjugates can be formed as well as the desired enzyme-protein conjugate. The nature of the products will depend on several factors (13): (a) the rate of reaction of the linking agent with solvent and the two proteins, (b) relative rates of reaction of agent with both proteins, (c) rate of intra- vs inter-molecular cross-linking, (d) relative size of both proteins.

Slow addition of the conjugation reagent is said to increase the yield (13). In general, yields of conjugate are low with the one-step glutaraldehyde reaction (9, 14) with less than 10% of immunoreactivity retained. This method cannot be recommended unless large quantities of material are available for conjugation.

Two-Stage Reactions

There are four main types of two-stage reaction (Table 2).

Table 2. Types of two-stage reactions

	Type	Example of reagent
1	Homobifunctional	Toluene-2,4-diisocyanate (TDIC)
2	Heterobifunctional	Toluene-2-isocyanate 4 isothio-cyanate
3	Differential react-ivity	Glutaraldehyde
4	Specific modifications	Periodate oxidation

In this type of reaction, one protein is reacted with the reagent, excess reagent is removed, and the second protein added to form the conjugate.

(a) Homobifunctional reagents

These are compounds that contain identical reactive groups of which toluene 2,4-diisocyanate (TDIC) is an example. The two isocyanate groups (NCO) are ortho and para to the methyl group on the benzene nucleus. The ortho isocyanate is sterically hindered by the adjacent methyl group and will react with amino, thiol and hydroxyl groups at 38°C, while the para group will react at 0°C (15). Thus the reaction need only be carried out at two temperatures. This reaction does not appear to have been used to form conjugates.

(b) Heterobifunctional reagents

Here, the reagent has two different reactive groups, e.g.

toluene 2-isocyanate 4-isothiocyanate. The isocyanate group is more reactive than the other and the first protein can be reacted with the reagent, which is then removed by dialysis before the addition of the second protein. This will minimize self-coupling by the first protein as the reaction conditions will not allow the isothiocyanate to react. This method is said to yield inferior results to TDIC (15) and does not appear to have been used to form conjugates.

(c) Reagents with differential reactivity

The only reagent in this group is glutaraldehyde, which is widely used to produce conjugates (16). The basis of the reaction is the different rate of reaction of glutaraldehyde with the enzyme and the protein. The system is limited to the use of one enzyme only, horse-radish peroxidase (HRP) (E.C. 1.11.1.7) which is relatively unreactive to glutaraldehyde owing to the blocking of free amino groups by the allyl isocyanates that occur in horse-radishes (17).

The enzyme is allowed to react with glutaraldehyde (18 hr. at room temperature) and filtered through a Sephadex G-25 column to remove unreacted glutaraldehyde. The activated enzyme is then reacted with protein (antibody) for 24 hours at $4^{o}C$, before terminating the reaction by excess lysine. Self-coupling of the enzyme is limited, and 1:1 enzyme-protein conjugates are said to be formed (18).

Comparison of the one-step and two-step glutaraldehyde methods shows the one-step method as yielding high molecular weight conjugates that are heterogeneous, while the two-step method yields homogeneous conjugates (18). When both methods were used to prepare HRP - human chorionic gonadotrophin (HCG) conjugates, the two-step method was found to be superior (19). Yields of conjugate were still low, usually less than 10% (14, 16). Using alkaline phosphatase (ALP E.C. 3.1.3.1) about 60% - 70% of the initial enzyme activity, and 1% - 10% of the initial immunological activity were retained in ALP-IgG conjugates (9) with half the enzyme activity bound to the immunoreactive IgG (20). Thus, the method, although widely used, is only suitable when large amounts of materials are available for conjugation.

(d) Specific modification to ensure selective reactions

In this type of reaction, one of the proteins is modified before reaction with the second protein. Provided certain conditions are met, little self-linking occurs, and the yield of conjugate is good. Three such methods have been described.

(i) N, N^1-o-phenylenedimaleimide (21). The enzyme β-galactosidase (E.C. 3.2.1.23) contains sulphydryl groups (-SH) which can be used in conjugation. The protein that is to be conjugated is reacted with mercaptosuccinic anhydride to form -SH groups. Alternatively, they may be generated by reduction of disulphide bridges (22). Unreacted material is removed by Sephadex G-25 chromatography and the treated protein is further reacted with N, N^1-o-phenylenedimaleimide solution and unreacted material removed by further Sephadex G-25 chromatography. The activated protein can then be reacted with the enzyme to form the conjugate. No enzyme activity is lost (21) and some 50% or more of the enzyme was attached to an IgG molecule (23). Little polymerization appeared to have occurred.

(ii) m-Maleimidobenzoyl-N-hydroxysuccinimide ester (MBS) (24). This reagent acylates amino groups, leaving the maleimide group free to react with thiol groups on the enzyme. To avoid polymerization, the protein must have no free thiol groups. The reaction is rapid and simple, with the protein and the MBS reacting for 30 minutes at room temperature. The product is harvested by centrifugation, washed and dried, dissolved in phosphate buffer and allowed to react with the β-galactosidase. The reaction conditions are mild, and at least 70% of the enzyme is linked to immunoactive protein. The loss of enzyme activity is small and the working assay (for insulin) is very sensitive (24).

The reagent was synthesized by the authors in the original

work, but is now commercially available (BCL, Lewes, East Sussex, U.K.). Similar compounds have been developed and are commercially available which can perform the same function. These are bis (malimido) methyl ether (BMME) (25) and malimidohexanoyl-N-hydroxysuccinimide ester (MHS) (26). Both reagents have been used to prepare conjugates, and yields and recoveries of enzyme and immunoactivity have been reported to be very good.

(iii) Periodate oxidation. This method of conjugation was briefly reported in 1973 (27) and expanded in 1974 (28). The method was developed to form conjugates for the intracellular localization of antigens, but has been used to form conjugates for use in enzyme-immunoassays (12, 29). The principle is as follows:

The enzyme, which must be a glycoprotein,is reacted with excess fluorodinitrobenzene (FDNB) to block free amino groups. The carbohydrate moieties are then oxidized with sodium periodate to aldehyde. The presence of FDNB or other low molecular weight compounds does not interfere with this stage (28). The oxidation is terminated by addition of excess ethyleneglycol, and the mixture dialysed overnight against a carbonate buffer, pH 9.5. The protein is then added and left for 24 hours at room temperature to allow free amino groups to react with the aldehyde to form a Schiff base. The reaction is stopped by the addition of ethanolamine, and the mixture is dialysed and chromatographed to yield the conjugate.

The reaction was modified (30) to eliminate the FDNB blocking step which did not entirely prevent self-conjugation of the enzyme. By carrying out the oxidation at an acid pH (4-5), self-coupling is prevented, and following dialysis against an acid buffer, the pH of the mixture is raised to 9.5 when the protein can be added. Following a two hour incubation, sodium borohydride is added and left for a further two hours before purification of the conjugate on a Sephacryl column. This will stabilize the Schiff base, enhancing conjugate stability (28). Although the enzyme can self-couple under alkaline conditions, only about 5% of the enzyme does so compared with 35% for the FDNB blocked enzyme (30).

The reaction can yield high (>70%) levels of conjugate (28) with good retention of immunological and catalytic activity. The reaction does need to be controlled carefully in order to obtain a 1:1 or 1:2 protein-enzyme ratio, as high molecular weight conjugates containing some 5-6 enzyme molecules per protein have poor immunoactivity (30). So far, peroxidase has been the enzyme of choice, but alkaline phosphatase has also been used to form a conjugate (12).

Enzyme-Hapten Conjugates

The commercial availability of kits for immunoassay of drugs and other haptens minimizes the need for synthesizing conjugates. However, assays have been reported for cortisol, oestriol and progesterone among others, so conjugates can be made if desired.

There are two important considerations to be noted. Firstly, the hapten may have to have a substituent group introduced so that it can further react with the conjugating reagent. Similar techniques are used to link haptens to carrier proteins, and will not be considered further.

The second point is the site of attachment of the hapten to the enzyme. If the same point is used to attach the hapten to the carrier protein and the enzyme, then the specificity of the assay may suffer (31). A second point of attachment should be sought.

Any of the reagents listed above can be used for conjugation. Thus, glutaraldehyde could be used to link a hapten containing an amino group and an enzyme, or carbodiimides to link carboxyl, amine, alcohol and thiol groups. Conditions can be adjusted to alter the degree of incorporation of the hapten, which can make a difference to the sensitivity of the assay. There is no easy way of determining the exact amount of incorporation required to give the most sensitive assay, and considerable experimentation may be required.

In general, three methods are in use for conjugation:

(a) Mixed anhydride

This has been widely used and is held to be the most satisfactory conjugation agent (16). The reaction involves reacting a carboxyl group on the hapten, either native or introduced, with secondary butylchloroformate. The mixed anhydride formed can be reacted with protein amino groups to form a peptide bond (32). Yields are good (32) and no enzyme activity is lost, provided the enzyme does not precipitate out (31).

(b) Carbodiimides

This is a two-step reaction in which the hapten is first reacted with the carbodiimide. The enzyme is then added, and excess carbodiimide removed by addition of glycine. When used to couple progesterone 11α-hemisuccinate to β-galactosidase, the recovery of enzyme activity was 90%, but only 25% of the preparation could bind to the antibody (33).

(c) MBS

This has been described elsewhere but would appear to be of use in forming conjugates.

Purification

This is essential to remove unreacted ligand as well as other reaction products. The unconjugated enzyme will raise the background activity in any assay where the free fraction is measured, while in the presence of free ligand or other fractions (e.g. polymers) can interfere with the sensitivity and precision of the assay (5).

Purification is usually accomplished by gel filtration on

Sephadex (12) or Sepharose (23) which can separate enzymes from protein ligands. Less common methods involve density centrifugation (5) and affinity chromatography (29). Similar methods are applicable to enzyme hapten conjugates, although separation of labelled enzyme from free enzyme can cause difficulties owing to the relatively small molecular weight difference between the free and conjugated enzyme. As the conjugation of enzyme and hapten can be easily controlled, reaction conditions can be adjusted to give the smallest amount of free enzyme after reaction.

Characterization

Comment has been made on the lack of attention paid to poor characterization of conjugates, most authors being content to report the development of a working assay (16). One group of workers (34) reported the characterization of alkaline phosphatase and lactoperoxidase antibody conjugates made by the glutaraldehyde method. The study was confined to gel chromatography investigation of conjugates formed at various concentrations of glutaraldehyde. Thus, they were able to define the conditions for producing conjugates. However, this type of study is rare.

Given that conjugation can interfere with the properties of the antibody and the catalytic activity of the enzyme, more care should be taken in the characterization of conjugates. The following parameters should be assessed:

(a) Enzyme activity

The effect of conjugation on the Michaelis constant (Km) and maximum velocity of reaction (Vmax) of the enzyme should be determined. Significant alterations in these parameters can lead to loss of enzyme activity and, hence, an insensitive assay requiring lengthy incubation times to develop a measurable colour. For EMIT assays a high level of activity is essential, as measuring times are short.

(b) Immunological activity

Conjugation can mask binding sites, antigenic determinants or sterically inhibit binding by the antibody. The reaction may also damage the molecule. The best test is the use of the conjugate in a working assay, and by performing an antibody dilution curve to check the conjugate.

(c) Proportion of tracer formed

This should be determined. A system that gives a low yield of working conjugate is wasteful, especially if only a small amount of ligand or binder is available.

Conclusion

The method of choice for conjugation should be quick, easy to perform, inexpensive and yield a high level of good quality conjugate. This state of affairs has not been reached yet, but some methods show promise. The periodate oxidation technique has been shown to give a good yield of working tracer (12) and not to significantly affect the Km or Vmax of alkaline phosphatase when conjugated to human placental lactogen (35). The procedure is not complicated, and does not require special apparatus or reagents. Similarly, the related group of compounds, MBS, BMME and MHS have been reported to give good yields of conjugate, and the reagents are available commercially and are easy to use. These methods are worthy of investigation and are commended to anyone who wishes to form their own conjugates. Glutaraldehyde, although widely used, tends to low yields of conjugate and should only be employed if large amounts of material are available for conjugation.

References

1. Rubenstein, K.E., Schneider, R.S., Ullman, E.F.: "Homogeneous" enzyme immunoassay. A new immunochemical technique. Biochem. biophys. Res. Commun. 47, 846-851 (1972).

2. Voller, A., Bartlett, A., Bidwell, D.E.: Enzymic immunoassays with special reference to ELISA techniques. J. clin. Path. 31, 507-520 (1978).

3. Maiolini, R., Masseyeff, R.: A sandwich method of enzyme immunoassay. 1. Application to rat and human alpha fetoprotein. J. Imm. Meth. 8, 314-317 (1973).

4. Masseyeff, R., Maiolini, E., Bouran, Y.: A method of immunoassay. Biomedicine 19, 314-317 (1973).

5. van Weemen, B.K., Schuurs, A.H.W.M.: Immunoassay using antigen-enzyme conjugates. FEBS Letts. 15, 232-236 (1971).

6. Kennedy, J.H., Kricka, L.J., Wilding, P.: Protein-protein coupling reactions and the applications of protein conjugates. Clin. chim. Acta 70, 1-31 (1976).

7. Mannix, M., Dawney, W.: Studies on the conjugation of horse-radish peroxidase to Fab fragments. J. Imm. Meth. 3, 233-242 (1973).

8. Avreamas, S.: Coupling of enzymes to proteins with glutaraldehyde. Immunochemistry 6, 43-52 (1969).

9. Engvall, E., Perlmann, P.: Enzyme linked immunosorbent assay (ELISA). Quantitative assay of immunoglobulin G. Immunochemistry 8, 871-874 (1971).

10. Engvall, E., Perlmann, P.: Enzyme linked immunosorbent assay (ELISA). III. Quantitation of specific antibodies by enzyme labelled anti-immunoglobulin in antigen coated tubes. J. Immunol. 109, 129-135 (1972).

11. Goodfriend, T.L., Levine, L., Fasman, G.D.: Antibodies to bradykinin and angiotensin. A use of carbodiimides in immunology. Science 144, 1344-1346 (1964).

12. Williams, D.G.: Enzyme immunoassay for human placental lactogen suitable for use in a routine hospital laboratory. In: "Enzyme Labelled Immunoassay of Hormones and Drugs", Ed. Pal, S.B., Walter de Gruyter & Co., Berlin, New York, pp. 129-136 (1978).

13. Modesto, R.R., Pesce, A.J.: The reaction of 4,4' difluoro-3,3' dinitro diphenyl sulphones with gamma globulin and horse-radish peroxidase. Biochim. biophys. Acta 229, 384-395 (1971).

14. Adams, T.H., Wisdom, G.B.: Peroxidase labelling of antibodies for use in enzyme immunoassay. Biochem. Soc. Trans. 7, 55-57 (1979).

15. Schik, A.F., Singer, S.J.: On the formation of covalent linkages between two protein molecules. J. biol. Chem. 236, 2477-2484 (1961).

16. Schuurs, A.H.W.M., van Weemen, B.K.: Enzyme immunoassay (Review). Clin. chim. Acta 81, 1-40 (1977).

17. Nakane, P.K., Sri Ram, J., Pierce, G.B.: Enzyme labelled antibodies for light and electron microscopic localization of antigens. J. Histochem. Cytochem. 14, 789-791 (1966).

18. Avreamas, S., Ternynck, T.: Peroxidase labelled antibody and Fab conjugates with enhanced intracellular penetration. Immunochemistry 8, 1175-1179 (1971).

19. van Weemen, B.K., Schuurs, A.H.W.M.: Immunoassay using antibody enzyme conjugates. FEBS Letts. 43, 215-218 (1974).

20. Engvall, E., Jonsson, K., Perlmann, P.: Enzyme linked immunosorbent assay II. Quantitative assay of protein antigen, immunoglobulin G, by means of enzyme labelled antigen and antibody coated tubes. Biochim. biophys. Acta 251, 427-434 (1971).

21. Kato, K., Hamaguchi, Y., Fukui, H., Ishikawa, E.: Enzyme linked immunoassay. I. Novel method for synthesis of the insulin-β-D-galactosidase conjugate, and its applicability for insulin assay.J. Biochem. 78, 235-237 (1975).

22. Kato, K., Hamaguchi, Y., Fukui, H., Ishikawa, E.: Coupling Fab fragment of rabbit anti-human IgG antibody to β-D-galactosidase and a highly sensitive immunoassay of human IgG. FEBS Letts. 56, 370-372 (1975).

23. Kato, K., Hamaguchi, Y., Fukui, H., Ishikawa, E.: Enzyme linked immunoassay. II. A simple method for the synthesis of the rabbit antibody-β-D-galactosidase complex and its general applicability. J. Biochem. 78, 423-425 (1975).

24. Kitagawa, T., Aikawa, T.: Enzyme coupled immunoassay of insulin using a novel coupling reagent. J. Biochem. 79, 233-236 (1976).

25. Weston, P.D., Devries, J.A., Wrigglesworth, R.: Conjugation of enzymes to immunoglobulins using dimaleimides. Biochim. biophys. Acta 612, 40-49 (1980).

26. Yoshitake, S., Yamada, Y., Ishikawa, E., Masseyeff, R.: Conjugation of glucose oxidase from aspergillus niger and rabbit antibodies using N-hydroxysuccinimide ester of N (4-Carboxycyclohexylmethyl)-maleimide. Europ. J. Biochem. 101, 395-399 (1979).

27. Kawaoi, A., Nakane, P.K.: An improved method of conjugation of peroxidase with proteins. Ed. Proc. 32, 840 (1973).

28. Nakane, P.K., Kawaoi, A.: Peroxidase labelled antibody. A new method of conjugation. J. Histochem. Cytochem. 22, 1084-1091 (1974).

29. Barbour, H.M.: Development of an enzyme immunoassay for human placental lactogen using labelled antibodies. J. Imm. Meth. 11, 15-23 (1976).

30. Wilson, M.B., Nakane, P.K.: Recent developments in the periodate method of conjugating horse-radish peroxidase (HRPO) to antibodies. In: "Immunofluorescence and Related Staining Techniques", Eds. Krapp, W., Holmbar, K., Wick, G., Elsevier/North-Holland Biomedical Press, Amsterdam, pp. 215-224 (1978).

31. van Weemen, B.K., Schuurs, A.H.W.M.: The influence of heterologous combinations of antiserum and enzyme labelled estrogen on the characteristics of estrogen enzyme-immunoassays. Immunochemistry 12, 667-670 (1975).

32. Erlanger, B.F., Borck, F., Beiser, S.M., Liberman, S.: Steroid protein conjugates. II. Preparation and characterization of conjugates of bovine serum albumin with progesterone, deoxycorticosterone and oestrone. J. biol. Chem. 234, 1090-1094 (1959).

33. Dray, F., Andrien, J.M., Renaud, R.: Enzyme immunoassay of progesterone at the picogram level using β-galactosidase as label. Biochim. biophys. Acta 403, 131-138 (1975).

34. Ford, D.J., Rodin, R., Pesce, A.J.: Characterization of glutaraldehyde coupled alkaline phosphatase-antibody and lactoperoxidase-antibody conjugates. Immunochemistry 15, 237-243 (1978).

35. Williams, D.G.: Comparison of three conjugation procedures for the formation of tracers for use in enzyme immunoassays. J. Imm. Meth. 72, 261-268 (1984).

THE MEASUREMENT OF INSULIN ANTIBODIES AND ITS INTERPRETATION

T. J. Wilkin
Department of Medicine II, University of Southampton,
Southampton General Hospital, Southampton SO9 4XY, U.K.

Introduction

Antibodies to insulin were first reported nearly thirty years ago in insulin-treated diabetics, when their influence on insulin kinetics was clearly demonstrated (1). Insulin resistance (2), allergy (3) and local lipoatrophy (4) have been the main clinical complications of insulin treatment, but their importance has declined over the years with the introduction of increasingly pure insulins. Most clinical studies of the past five to ten years have focussed on trials designed to measure the antibody response to various formulations of the new monocomponent insulins (5, 6). Curiously, even human insulin has been shown to provoke an antibody response in up to 14% of diabetic patients (7).

More recently, attention has turned to the phenomenon of autoimmunity to insulin. Autoantibodies to insulin in newly diagnosed but untreated diabetic children have been described (8). Subsequently, insulin autoantibodies were identified in 37% of newly diagnosed type I diabetic adults and in 48% of healthy co-twins of type I diabetics (9). Some autoantibodies which are specific for human insulin are not detectable using porcine or bovine ligands (10).

Most diabetic sera contain insulin antibodies which bind human, porcine and bovine insulins more or less equally, although bovine may predominate (11). Little clinical interest has been shown in the epitopes involved and such responses have been

Immunoassay Technology, Vol. 2

looked upon as polyclonal and largely unselective. Data currently being prepared, however, suggest that while antibodies provoked by heterologous insulins tend to select A-chain determinants, insulin autoantibodies commonly favour (sometimes exclusively) sites on the B chain (12). Specificity of determinant may, therefore, be of considerable theoretical and even practical importance in the interpretation of binding reactions.

A wide variety of assays are available for the measurement of insulin antibodies. Most are liquid phase systems based on the Farr assay, in which label bound to antibody at equilibrium is separated from free label by phase partition, electrophoresis, or adsorption. Solid phase assays provide a means of retaining complexed antibody within the system, and enzyme-linkage obviates the need for radiochemical tracers. All assays are liable to be influenced to a greater or lesser extent by insulin in the test serum and by the relative quantities of antibody and antigen in the assay. Polyclonal sera are antibody mixtures, and interpretation of binding data should be appropriately restrained.

This review outlines the properties of insulin antibodies and examines their binding characteristics in different assay systems. Special attention is paid to enzyme-linked immunosorbent assay (ELISA) and to the measurement of autoantibodies to insulin.

Types of insulin

Insulin is a polypeptide of molecular weight 5,600 and of known tertiary structure (14). The molecule comprises an A chain of 21 amino acids and a B chain of 30 residues. There are no repeat sequences in its structure. The chains are joined by two disulphide bridges. A third disulphide bond between A6 and A11 forms a pleat in the A chain known as the α-loop. Certain areas of both chains, principally the cysteine residues to which the disulphide bonds attach, and the site of biological activity (B22-26), are evolutionarily stable. Other areas, the α-loop particularly, show major differences between species.

Beef, porcine and human insulins are the variants most commonly used in clinical practice. Porcine insulin differs from human by the substitution of a single residue at B30, and bovine by the residue at B30 and two others in the α-loop. All three variants share similar biological activity.

Insulin is produced from the enzymatic cleavage of pro-insulin in the Golgi apparatus of islet B-cells. The C-peptide liberated during proteolysis shows much greater species variation (and antigenicity) than insulin. Bovine C-peptide, for example, differs from the human molecule in 14 of its 31 residues (15). Insulin is stored in B-cell vesicles as a hexameric crystalline aggregate, but circulates as the free monomer (16).

The formulation of insulin for clinical use has greatly improved over the years. Early preparations of bovine and porcine insulins contained not only proinsulin and C-peptide contaminants, but immunologically significant quantities of other pancreatic hormones as well (17). Historically, insulins were purified by repeated recrystallisation, but the use of chromatography has led to essentially monocomponent preparations (18). Recent technical advances have led to the availability of biosynthetic (19) and semi-synthetic (20, 21) human insulins on a commercial scale. Therapeutic insulins are variously formulated to provide different durations of action. Retarding agents such as fish protamine are used to bind insulin (22) and zinc to encourage polymerisation (23).

Antigenicity of insulin

Theoretically, the immune response to injected insulin should be limited to foreign residues (i.e. to ALA-B30 in the human response to porcine insulin). This is not observed to be the case, however, (24, 25) and responses to determinants shared by both host and donor can probably best be explained by a carrier effect, either of insulin polymers (see below) or of proinsulin, if present, as a contaminant (25). The greater antigenicity of bovine insulin sometimes seen in humans (11) has been ascribed to the α-loop differences, but again may result from contaminating proinsulin which is present in greater concentration in bovine than in the highly purified porcine insulins (26).

Species differences and formulation apart, insulin antigenicity depends upon mode of administration (27), as well as numerous host factors (28). Insulin given intermittently (e.g. during recurrent gestational diabetes) appears to be far more antigenic than regular doses, which may even induce tolerance (29). The observation that insulin prepared from a particular species may be antigenic for the same species (24, 25), suggests that it is the subcutaneous mode of administration which renders the molecule antigenic. Endogenous insulin circulates as a dilute low molecular weight monomer (30), whereas insulin injected subcutaneously remains for long periods in high concentration at 37°C, tending to encourage polymer formation. Furthermore, there are always small amounts of non-dissociable higher molecular weight polymers of insulin, even in highly purified preparations (27). Polymerisation may simply provide the concentration of carrier determinants necessary to stimulate T-cell activity, or it may influence molecular shape. Even dimerisation is known to distort the tertiary structure of insulin (31). The addition of protamine (32) or zinc (33) to long-acting preparations is likely further to increase the antigenicity of injected insulin.

Autoimmunity to insulin has only recently been recognised (10, 34). Here, the antigen presumably circulates as the free monomer, and every determinant can be regarded as a monovalent hapten. The α-loop may be an important carrier site (35).

The immune response to insulin

Immunoglobulin structure is based on the IgG monomer which assumes the form of a letter "Y". The monomer has two antigen binding sites (idiotopes) on the arms of the "Y" and a biologically functional region on its stem. IgM is pentameric and

IgA either monomeric or polymeric with respect to IgG. Antigens such as the pneumococcus polysaccharide have appropriately spaced repeating structures capable of stimulating B lymphocytes directly, without the need for T-cell cooperation. When thus stimulated, B lymphocytes tend to secrete IgM which, by virtue of its 10 identical idiotopes, is highly avid for the repeat structures of the polysaccharide (36). Insulin, in contrast, is a small, asymmetric molecule without repeat structures. The B lymphocyte response to any one of its haptens will depend on the rest of the molecule or polymer providing carrier function adequate to active T-cells, and the product is mainly IgG antibody. The more efficient the carrier, the greater the activation.

Impure or polymerised preparations of heterologous insulin delivered subcutaneously provide substantial carrier function (27), and antibodies to a wide variety of haptens result (37). Poorly antigenic substances such as circulating insulin monomers, however, provide little carrier function, and even if antibodies occur through a defect in immune tolerance (i.e. autoimmunity), their specificity is likely to be restricted to a few immunologically dominant haptens (38). Insulin antibodies (3, 39) are predominantly IgG, although the author (unpublished) and others have been able to detect low titres of IgA and IgM in some sera.

Several studies suggest that HLA haplotypes influence immune responsiveness to heterologous insulin (40, 43). Results have not been consistent, however, and have recently been challenged in a prospective study which shows no significant relation between HLA and reactivity to therapeutic insulin (44). Clinical studies such as these have, nevertheless, looked at overall insulin binding capacity; they do not negate the concept, derived largely from animal studies, that determinant selection in antibody response is an HLA related phenomenon (45).

Circulating insulin/antibody complexes are small and probably monomeric (46), due on the one hand to the lack of repeating structures to which idiotypically similar IgG molecules might otherwise bind and, on the other, to the small size of the antigen, causing stearic hindrance to the binding of molecules of different idiotype. Polymerisation of complexes may occur where antibody concentration is very high (47, 48).

The fact that sera from insulin-treated diabetics will usually bind human, porcine and bovine insulins equally, suggests that shared epitopes are involved. In our experience, diabetic sera will also bind desalinated pork insulin (B30 removed, and with it the single residue distinguishing pork from human insulin), showing how clones in this response are activated against self haptens.

Autoantibodies to insulin show limited clonality (12). While some are complementary to A chain epitopes shared by several variants, others are specific for the THR-B30 residue and bind neither porcine nor bovine, but only human insulin (10). Occasionally, insulin antibodies bind so much insulin that its inappropriate release results in hypoglycaemia (34). This phenomenon was first reported in an insulin-treated diabetic (49); however, we have recently documented a similar situation in an insulin-naive individual with autoantibodies specific to human insulin (for publication).

The detection of insulin antibodies

The detection of specific antibody depends on its binding with pure antigen. Whether the reaction takes place in liquid phase, gel or on solid phase, it is necessary to separate free from bound elements and measure one as a proportion of the total, the other or an independent reference.

Assay systems are traditionally classified into solid/liquid phase, competitive/non-competitive, radio/non-radiochemical etc. More fundamental, however, are the relative concentrations of reactants within the system, there being basically two situations: relative antigen excess and relative antibody excess. The thermodynamics of each are likely to be very different. Antigen valencies tend to be saturated in antibody excess (e.g. liquid phase radioassay in which the ligand is a trace quantity of high specific activity) giving a molar ratio Ab:Ag. Antibody valencies on the other hand tend to be saturated in antigen excess (e.g. solid phase systems where the ligand is immobilised) giving a different molar ratio, Ag_2:Ab. We have calculated for example (see later) that an average 48 ng insulin binds to the solid support in the optimised ELISA which we employ. In the only liquid phase radioassay sufficiently analysed in the literature (50), the optimum quantity of labelled insulin was found to be 125 pg (lower by a factor of 384). The insulin in a solid phase system is, moreover, distributed in two rather than three dimensions, so that the density of insulin molecules presented to the serum under test is likely to be many thousands of times greater. Reactant ratios are particularly important when the antibodies in test sera are polyclonal, displaying a hierarchy of affinities. Antigen is attracted most strongly to the antibody of highest affinity; when this is saturated, remaining antigen is free to equilibrate with the antibody of next highest affinity, and so on. This has two practical implications: when antigen is limited relative to antibody, some of the lower affinity antibody under test may not be represented at all in the reaction and secondly, while the quantity of antigen bound will rise with increasing antigen added, the percentage of antigen bound will fall as antibody of lower and lower affinity is recruited (50). Antigen excess assays should normally permit access to binding of all specific antibody, each clone reaching its appropriate equilibrium independently.

Sera for insulin antibody estimation normally contain endogenous free and, if antibodies are present, bound insulin. The quantity of bound insulin increases with antibody titre. This insulin cannot be ignored, because it will compete for antibody receptors and give a lower binding result than if the insulin had been removed. It may be argued, however, that a measure of free antibody binding sites is clinically more meaningful than total binding sites, and some contend that removal of insulin does not materially affect the interpretation of results (27). In theory, antigen excess assays should be less influenced by bound insulin than antibody excess systems, where the competition between large amounts of endogenous insulin and trace quantities of labelled ligand may be overwhelming. Circulating insulin-receptor antibodies (51), or anti-idiotypes to antiviral antibodies (52) may be expected to interact in the same way as endogenous insulin.

Separation of free from bound antibody is achieved by aspiration in solid phase systems, and by precipitation of complexes, electrophoresis or absorption of free ligand in liquid phase (53). Precipitation in liquid phase systems is achieved either physico-chemically by polyethylene glycol or a similar substance, or immunochemically by means of precipitating antibody. The former is simple but non-specific, the latter more demanding but advantageously immunospecific. Being mono - or at best dimeric (46), insulin containing immune complexes are highly soluble and do not form spontaneous lattices, even at equivalence. The absence of lattices reduces the precipitating efficiency of second antibody, which becomes entirely dependent for precipitation on widespread cross-binding of the Fc portions of IgG molecules. Insulin complexes may then comprise only a small proportion of the total IgG precipitated by second antibody, reducing signal strength and sensitivity.

Optimum binding of antibody in a solid phase system depends on obtaining an antigen monolayer of ideal density (54). A bilayer causes reduced binding (presumably as a result of stearic hindrance) and loss of precision probably through elution of antigen (55). Other considerations in the optimisation of solid phase assays, with particular emphasis on ELISA, are explored in more detail later in this article.

Units of Measurement

The binding unit in immunology is the bond between epitope and idiotope. The strength of binding is referred to as affinity.

The idiotopes on the two arms of an IgG molecule are the same, and identical with those of other molecules from the same clone. The binding units of a monoclonal serum are therefore homogeneous. Each has the same affinity for antigen, and the binding capacity of the serum - within the constraints of saturation and stearic hindrance - reflects the number of immunological bonds present. The Scatchard plot of a monoclonal serum is linear because the percentage of antigen bound does not change with the mass of antigen bound. Put alternatively, the ratio between the forward reaction constant and backward reaction constant, i.e. the affinity constant, is the same for all bonds and therefore the same at all concentrations of antigen. Scatchard analysis can quantify antigen/antibody reactions in absolute terms, but only if the test serum is monoclonal for the antigen.

The majority of test sera, however, are polyclonal (2), with a binding hierarchy determined by the affinity constants of the antibodies from each clone. Scatchard plots for polyclonal sera are curvilinear owing to the variety of bonds (56), and the percentage of antigen bound varies inversely with the quantity of antigen bound (53). There is no justification for speaking of high and low affinity populations of insulin antibody in a serum when the continuous curve of its Scatchard plot indicates broad heterogeneity. Neither is there any justification for expressing polyclonal binding as mass of labelled insulin per unit volume of serum, since the result may be shifted dramatically by altering the quantity of labelled insulin in the assay (53). Absolute units give a false image of accuracy to what is an arbitrary measure.

Less information is available regarding the effects of reactant concentrations on the stability of measurements from solid-phase assays for insulin antibody. It might be predicted, however, that binding of antibody would be independent of reactant concentrations provided antigen was homogeneous and presented in large excess. Results of antigen excess assays may be expressed as the raw signal, as end-point titre (58), or as ratio between the signals obtained from test serum and a reference (e.g. "ELISA index", 59) or negative serum. An alternative is to incorporate a multi-point reference curve in each assay run and express a result as the percentage dilution of the reference serum giving the same signal as the test sample (10).

The aim of units is to provide a basis for comparison between two measurements. The use of units assumes that like is compared with like. Difficulties in interpretation arise, however, when comparing binding capacities of two sera, one of which, for example, binds only insulin B chain while the other binds predominantly A chain. Such sera exist (see later), mainly in the context of insulin autoimmunity, and serve to emphasize the problems of attributing absolute units to uncharacterised antibody binding reactions. This problem is clearly least when the test sera are polyclonal, but the clonality of insulin antibodies is not predictable (12). Taking these problems of interpretation together, the best approximation to a satisfactory unit for expressing antigen binding by a polyclonal serum would seem to be a percentage unit or ratio applied under carefully standardised assay conditions.

Assays in current use

Insulin is a non-precipitin, and until the introduction of the radioisotope tracer method of demonstrating insulin/gamma globulin complexes (1), there was no direct evidence for antibodies to insulin in human serum. It was observed in 1958 that free I-131 labelled bovine serum albumin (I*BSA), which is also a non-precipitin, was soluble in 50% saturated ammonium sulphate, whereas I*BSA bound to antibody was not (13). Ammonium sulphate, and substances with similar properties (e.g. polyethylene glycol), are now widely used in radioassays to precipitate otherwise soluble immune complexes such as insulin IgG. An alternative, the use of immunospecific anti-human gamma globulin to precipitate labelled complexes was also developed (60). This approach was improved by optimising the precipitation of complexes and by using a volume marker to obviate the need for washing the separated precipitate (50). Phase separation has also been achieved by charcoal absorption of the free ligand, gel filtration, cellulose and ultracentrifugation.

Quantitative (rocket) immunoelectrophoresis to measure insulin antibodies has also been described (61). The test serum, containing insulin antibody preincubated with radio-insulin, was made to run electrophoretically through an agarose gel impregnated with anti-human IgG. The immune precipitate so formed was removed by cutting out the gel, and its radioactivity interpolated against a reference curve to give the amount of insulin IgG antibody present in the sample. In the light of observations that the reactivity of insulin with insulin antibodies was variable following iodination of the ligand (62, 63), a liquid-phase radioassay was designed in which the antibodies in test sera were made to compete for tracer ligand with precipitated guinea-pig anti-insulin antibodies (64). The partition ratio of label between test and precipitated antibody was felt to be independent of ligand affinity.

Despite the wide variety of assays described, little attention has been given to their characterisation. Attempts to remove insulin from the sera prior to testing have been made in only very few studies. It was claimed, in one study, that the presence of insulin did not materially alter the assay result (64), while in another, a sizeable error consequent on failing to remove insulin was demonstrated (50). The radioimmunoelectrophoretic method (61), and the radiobinding assay (50), are the only two to use immunospecific phase separation. The use of anti-human immunoglobulin to separate complexed insulin provides important direct evidence that the binding factor is an antibody.

Solid phase assays have been less widely used for insulin antibody determination, but are nevertheless well described. Thus, plastic tubes (54), paper discs (65) and sepharose beads

(66) have all been investigated as antigen supports, principally for IgG determination. The factors influencing an enzyme-linked assay for IgG insulin antibodies carried out in plastic tubes, were analysed (54). The assay compared favourably in terms of specificity, sensitivity and convenience with a radiobinding assay using polyethylene glycol as phase separator.

A Micro ELISA for Insulin Antibodies

The assay method depends on the detection of insulin antibody, retained by the insulin coated wells of a microtitre plate, by means of an enzyme/anti-human immunoglobulin conjugate which induces quantitative colour change in a substrate.

Preparation of peroxidase/anti-human Ig conjugate

Horseradish peroxidase (Sigma Chemicals, Poole) was conjugated to rabbit anti-human Ig (Dako, Mercia Brocades, Weybridge, Surrey) by the periodate method (67). The specificity of the conjugate for human Ig was demonstrated by substituting rabbit for human serum as a primary coating protein in a two-step ELISA, and PBS-Tween for test serum in the insulin antibody assay described below. The concentration of conjugate was optimised by trial and error to obtain maximum binding range.

ELISA procedure

The principle is shown in Fig. 1., 3.0 µg of highly purified human, porcine or bovine insulin (kindly donated by Eli Lilly Company and Novo Industries) diluted in 200 µl of 0.05 M carbonate-bicarbonate buffer of pH 9.6 were incubated under cover for three hours at 4°C in 94 wells of a 96-well polystyrene microtitre plate (Kontron Instruments Ltd., St. Albans, U.K.). Excess was removed and the plate washed three times by addition and aspiration of 0.15 PBS-Tween buffer of pH 7.2 (sodium chloride 8g, potassium chloride 0.2g, disodium hydrogen phos-

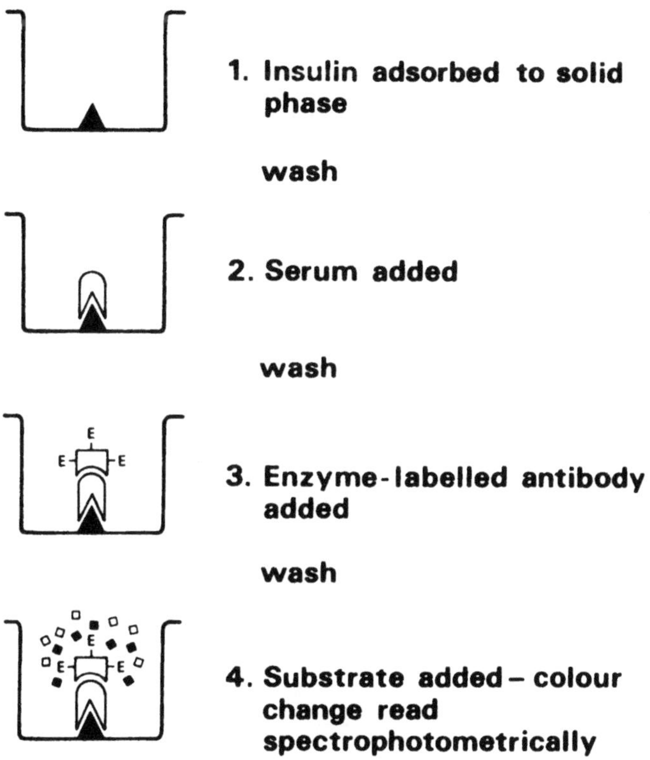

Fig. 1. The principle of a direct ELISA for the measurement of insulin antibodies.

phate 1.15g, potassium dihydrogen phosphate 0.02g, Tween-20 0.5 ml and gelatine 0.5g in 1 litre distilled water) and left empty.

The test wells were then filled with duplicate 200 µl samples of serum diluted 1:n in PBS-Tween buffer and incubated for two hours at room temperature. The wells were again flushed three times with PBS-Tween buffer, filled with 200 µl of the peroxidase-anti-human IgG conjugate diluted 1:1000 in PBS-Tween buffer, and left overnight at 4°C.

The substrate was prepared by placing 34 mg O-phenylene-diamine (Sigma Chemicals) in a dark glass bottle and adding 100 ml freshly made 0.15 M citrate-phosphate buffer, pH 5.0 (one part citric acid 21 g/l and one part disodium hydrogen phosphate 35.6 g/l) followed by 10 µl hydrogen peroxide, 30% w/v. The plate was emptied and in subdued lighting 200 µl of the substrate were added to all 96 wells. The plate was incubated in the dark for 30-40 minutes at room temperature and the reaction stopped by the addition to each well of 200 µl 34% sulphuric acid.

The light absorbance readings were made on a Dynatech MR 580 Micro Plate Reader (Dynatech Laboratories, Billingshurst, U.K.) at wavelength 490 nm, blanking with 200 µl substrate in the two unused wells.

The results are expressed as percentages derived from a reference curve for the appropriate species of insulin, obtained from serial dilutions of a diabetic serum and incorporated into every microtitre plate. Each result is a dilution equivalent, i.e. the percentage dilution of the reference serum (100% = 1:n) giving the same colour change as the test serum diluted 1:n. Owing to the high antibody content of some sera, a few values exceeded 100%. (The dilution factor, n, varied from 1:10 to 1:50 according to circumstances).

The variation in coating protein adhering to microtitre wells

All 96 wells of a microtitre plate were filled with 3 µg purified human insulin in 200 µl carbonate/bicarbonate buffer, to which had been added 56,000 cpm (56 ng) human insulin labelled by the iodogen method (68). The ^{125}I insulin retained by each well after incubation for 3 hours at 4°C is shown in Table 1, and averaged 1.6% of the insulin added (i.e. 49 ng). Binding was considerably lower in the edge wells, particularly the corner wells with two sides rather than one exposed. The "edge-effect" of microtitre plates is well known, although binding is usually higher in the peripheral wells (69, 70).

Table 1. The ^{125}I human insulin (cpm) retained by the wells of a microtitre plate after incubation of 56,166 cpm (56 ng) of labelled human insulin with 3 µg non-radioactive human insulin in 200 µl carbonate-bicarbonate buffer for 3 hours at 4°C.

	1	2	3	4	5	6	7	8	9	10	11	12
A	450	864	698	834	641	670	779	705	807	845	809	662
B	516	738	763	839	771	836	880	963	966	999	733	720
C	521	503	690	975	822	1033	1039	920	1089	951	751	708
D	471	571	824	1027	991	933	1075	1094	705	899	856	580
E	611	830	1149	1210	1034	1022	867	1098	1201	978	713	551
F	594	908	1140	1268	1159	1135	1136	1197	844	931	789	530
G	583	906	1179	1360	1311	1250	1278	1078	984	769	786	538
H	394	734	977	1051	1190	1251	1186	1124	1055	978	831	489

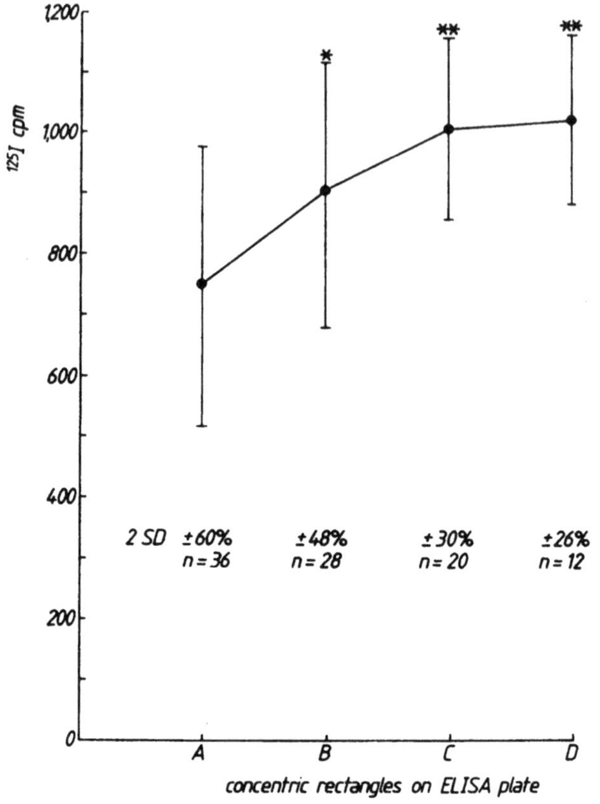

Fig. 2. The mean $\pm$ standard deviation of counts remaining in the wells of a microtitre plate following incubation with 56,000 cpm (56 ng) ^{125}I insulin added to 3 µg highly purified insulin in 200 µl buffer/well. A=perimeter wells, B=the concentric rectangle of wells internal to A, C internal to B and D internal to C. The mean values for B, C and D were significantly higher than those for A.

Fig. 2 shows the mean and variation ($\pm$ SD) of the % insulin bound in the peripheral, intermediate and central wells of the plate. The peripheral wells showed statistically significantly lower binding than the others. The variation in the peripheral wells was ± 60%, falling to ± 26% in the central block. A study was later carried out to establish to what extent this variation affected the final signal.

Variation in signal strength with concentration of coating protein

The variation in signal strength obtained by varying the concentration of coating insulin in an otherwise standardised assay (reference serum 1:20, conjugate 1:1000) is shown in Fig. 3. The relatively broad plateau means that the wide variance in insulin adhesion described above need make little difference to the final signal provided a judicious choice in concentration of coating protein is made. Thus the signal variation associated with placing 3 μg insulin in each well was very low. When attempts were made to economise on highly purified insulins by placing 0.3 μg insulin in each well, the coefficient of inter-assay variation rose dramatically to 15%-20% (not shown), and such economies were abandoned.

Extraction of insulin from test sera

Insulin was extracted by a modification of Dixon's method (71): To 2 ml undiluted test serum in a 10-ml polystyrene tube was added sufficient 0.85 M HCl to reduce the pH to 2.5. One ml of a thoroughly stirred dextran/charcoal (dextran 30 g/l-Norit GSX 120 g/l) suspension in methylcellulose (1 g/l in PBS - albumin buffer, pH 7.2) was added and mixed for five minutes on a rotary roller. The acid was used to dissociate insulin from its antibody, and the charcoal to adsorb it. The pH and ionic strength were immediately restored by the addition of 2.5 ml sodium phosphate buffer, pH 11.3 and the mixture adjusted to pH 7.4 with 0.85 M HCl as required. The tube was centrifuged at 2,500 rpm for 10 minutes and the supernatant taken to a dilution equivalent to 1:n of the original serum by the addition of assay buffer.

In order to check completeness of the extraction procedure,

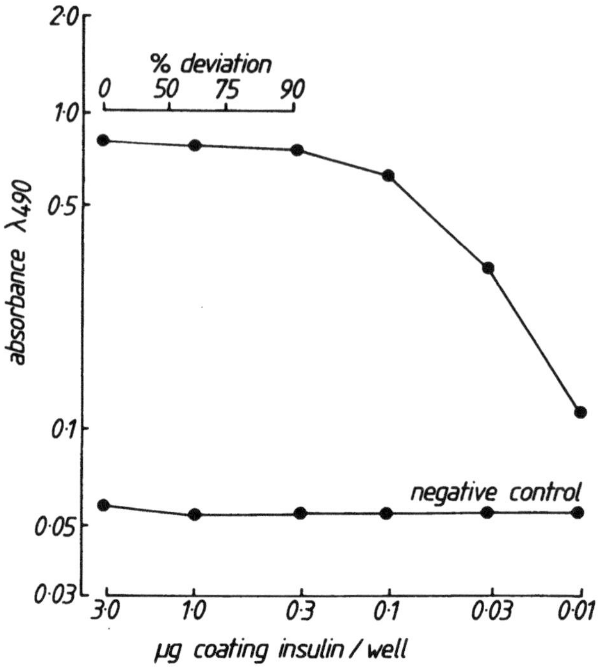

Fig. 3. The ELISA signal obtained by incubating graded quantities of insulin (3 µg to 0.01 µg) in the wells of a microtitre plate (peripheral wells not used). The deviation scale is added to facilitate interpretation of the data with reference to Fig. 2.

25 µl of iodogen-labelled ^{125}I insulin were added to the sera of nine insulin-treated diabetics and allowed to incubate overnight at 4°C and one hour at room temperature before commencing the extraction. The activity retrieved in the supernatant was measured as a percentage of the total counts added. The nine extracted diabetic sera which had been treated with labelled insulin retained a mean of only 3.16 SEM $\pm$ 0.2% of their original insulin content, confirming the efficiency of the extraction procedure. The correlations in insulin binding between 53 pairs of unextracted and extracted sera tested against human,

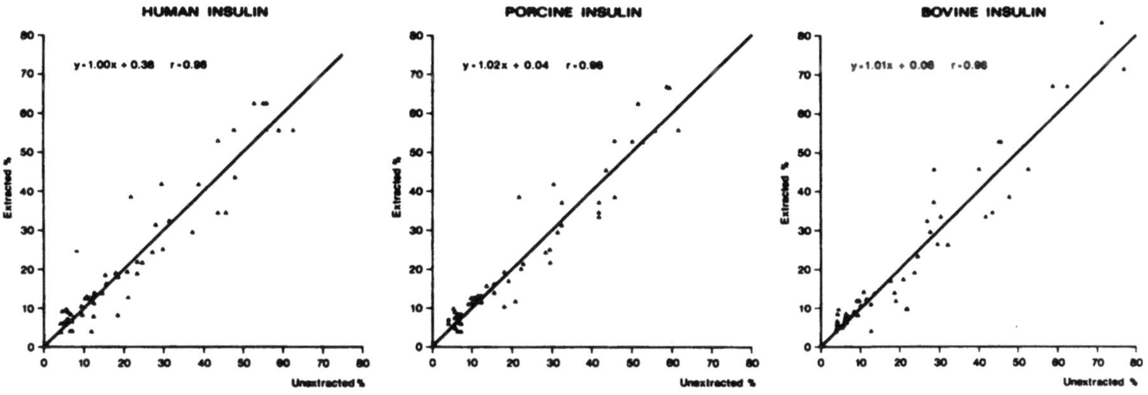

Fig. 4. ELISA: results of insulin binding (% dilution reference serum) obtained before and after extraction of insulin from 53 diabetic sera (identity: y = x).

porcine and bovine insulins are shown in Fig. 4. The regression equations indicate virtual identity (y = x) with correlation coefficients of 0.95 - 0.96. The results imply that prior extraction of insulin is not necessary using an ELISA for insulin antibodies. The dispersion of individual points probably reflects random error in the extraction process; error due to the presence of endogenous insulin would have been systematic. The range of binding capacity covered is typical of that seen in current diabetes practice.

Reference curve, precision and specificity

Reference curves are shown in Fig. 5 and inter-assay precision data in Table 2. The inter-assay coefficients of variation (standard deviation/mean x 100), based on 10 assays performed independently on different days, were 10% or less at widely spaced points (A and B, Fig. 5) for all three insulins. Corresponding coefficients for intra-assay precision were <4%. Control sera were drawn from 94 healthy adults (mean age 38 years, SD ± 7 years) with no history of diabetes or contact

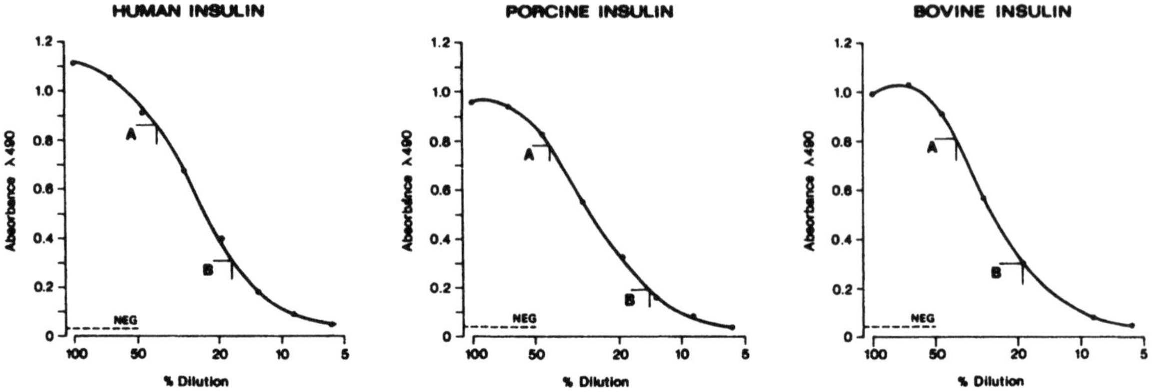

Fig. 5. Reference curves for insulin antibodies using human, porcine and bovine insulin, obtained from serial dilutions of serum from an insulin-treated diabetic. A and B represent the binding of quality control sera.

with insulin. The pooled normal sera gave an absorbance value (λ 490) of 0.042, equivalent to a <5% dilution of the reference serum. The range of binding of normal sera was 0.000-0.098 (λ 490), equivalent to 0 - 10.2%.

Table 2. Inter-assay coefficients of variation (%) at widely spaced points (samples A and B, see Fig. 5) on reference curves for the binding of serum from an insulin-treated diabetic to highly purified human, porcine and bovine insulin.

Insulin	Sample A (n = 10)	Sample B (n = 10)
Human	9.6	10.0
Porcine	8.6	9.8
Bovine	9.0	8.7

Results obtained by substituting purified human γ-globulin (Cohn fraction II; Sigma, Poole, U.K.) rich in IgG for test serum in the ELISA system were uniformly negative (<5% binding). Specificity for insulin, tested by substituting pure glucagon (Novo Industries 3 μg well) as coating protein, was essentially complete (Table 3).

Table 3. Dilution curve of reference serum applied to plates coated either with human insulin or with glucagon. Results in light absorbance units (λ 490).

Conc. reference serum %	100	67	44	29	20	13	8.8	5.9	3.9	0
Insulin binding λ 490	0.73	0.69	0.64	0.39	0.27	0.18	0.11	0.07	0.04	0.00
Glucagon binding λ 490	0.03	0.02	0.02	0.01	0.01	0.01	0.00	0.00	0.00	0.00

Inhibition studies

Attempts to inhibit antibody binding immunospecifically were made by adding increasing amounts of non-radioactive highly purified human, porcine and bovine insulin to known antibody positive and antibody negative sera, and preincubating overnight at 4°C. The results (Fig. 6) suggest dose-dependent inhibition by all three species variants, with no influence on the negative serum.

Relative Sensitivities of RIA and ELISA

Four unextracted sera from diabetics treated with insulin, and four from insulin-naive non-diabetics containing autoantibodies to insulin were serially diluted from 1:20 to 1:5,120 in assay

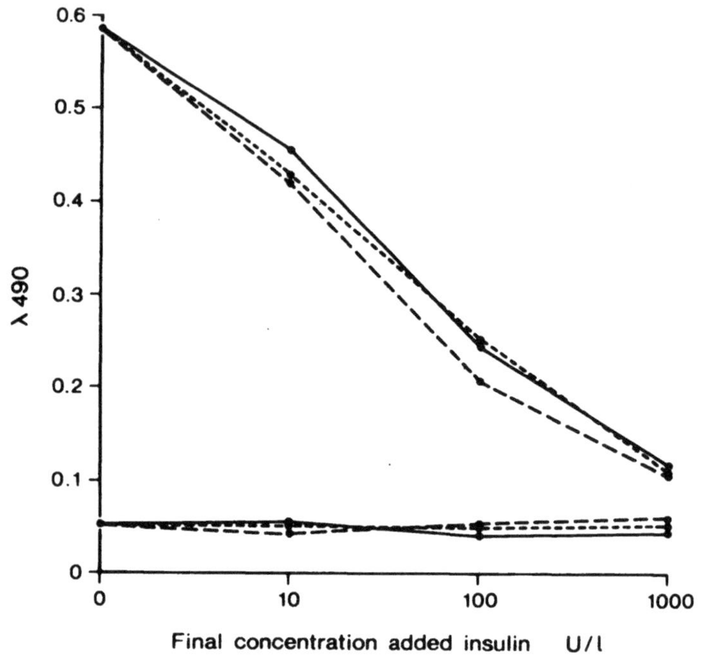

Fig. 6. Inhibition curves showing the relationship in insulin antibody positive and negative sera between binding to human (——), porcine (---) and bovine (···) insulin, and concentration of non-radioactive insulin added to the incubate.

buffer and tested against human insulin in the ELISA and in a radio binding assay. In addition, the sera of ten normal individuals and of the nine insulin autoantibody-positive subjects reported earlier by this group (10) were tested in both assays at a dilution of 1:20.

The radioassay was carried out as follows: to 500 µl of appropriately diluted serum in PBS were added 25,000 cpm human insulin (specific activity 1000 cpm/ng) in a plastic tube. After incubation overnight at 4°C, the bound label was precipitated by the addition of PEG to 12% final dilution, centrifuged and washed three times with PEG. The mean ± standard deviation of binding from duplicate assays of 10 normal sera was 1.6 ± 0.48%.

The results of binding (Fig. 7) show the ELISA to be more sensitive to insulin antibodies than the RIA. The RIA was sensitive to the strongest diabetic serum at 1:130, the ELISA at 1:640. The difference was more striking for insulin autoantibodies, where the RIA was sensitive to the strongest serum at 1:240, the ELISA at 1:5,100. Two of the autoantibody sera, giving undetectable signals in the RIA at all dilutions, were readily detectable in the ELISA.

Table 4 records the results of assaying the 19 individual sera at a dilution of 1:20. The nine insulin autoantibody containing sera, many of them human specific, were clearly detectable in the ELISA, but negative or barely detectable in the RIA. This was of particular interest as these nine sera had

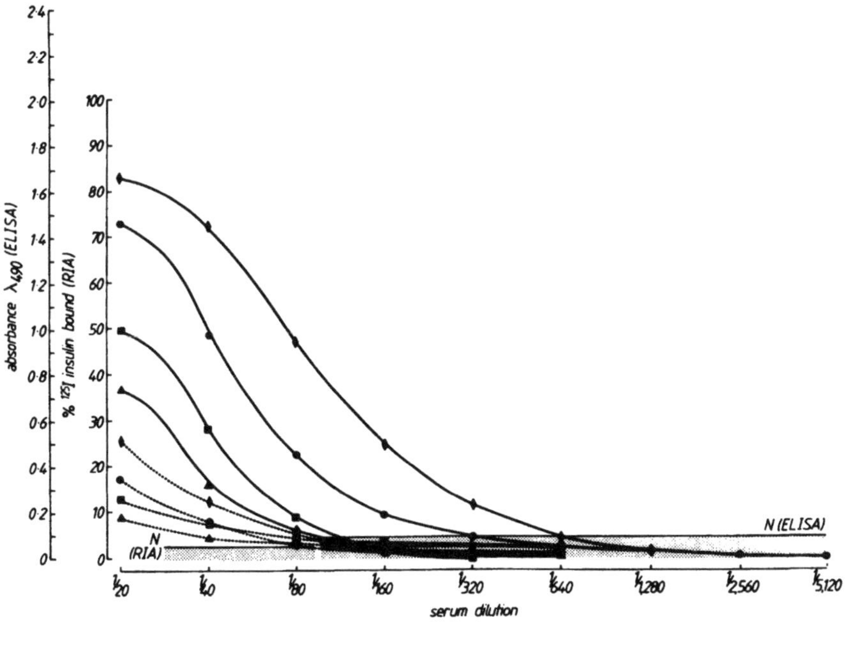

Fig. 7 - Upper panel

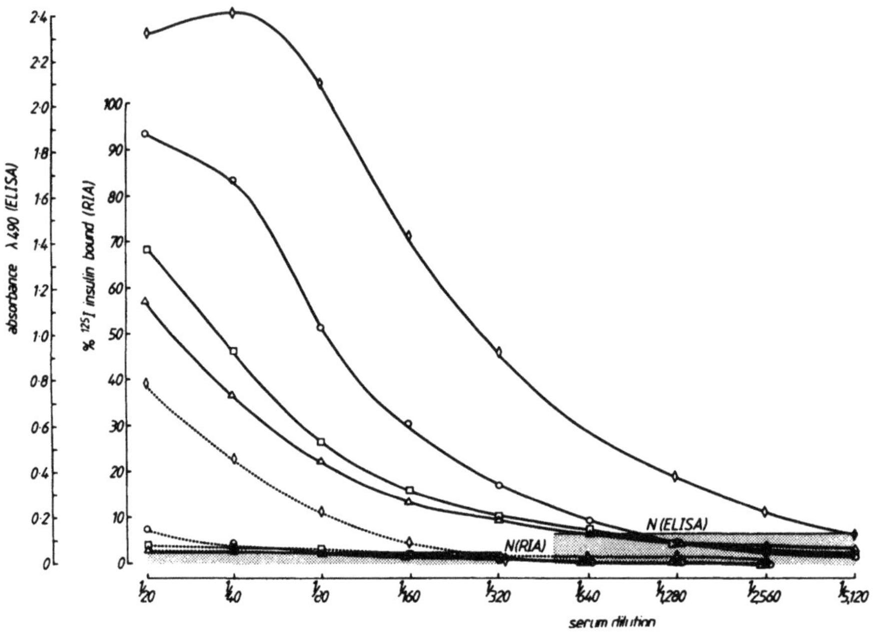

Fig. 7 - Lower panel

Dilution curves for insulin antibodies against human insulin obtained from the sera of four insulin-treated diabetics (upper panel) and four insulin autoantibody positive non-diabetics (lower panel) using ELISA (——) and RIA (---). The shaded areas indicate the upper limits of binding for the same ten negative control sera in each assay.

been deemed essentially negative for autoantibodies by three independent centres in the U.K. and Europe, all using liquid-phase radiobinding assays.

Correlation of ELISA with RIA

The ELISA was compared with an optimised immunospecific radio-

Table 4. Binding to human insulin of sera (at dilution 1/20) drawn from the nine subjects with autoantibodies to insulin described previously (10).

Patient	RIA binding %	ELISA absorbance λ 490
DP	1.6	0.42
KW	2.8	0.91
AP	3.4	1.30
JAv	2.3	1.30
JAr	3.6	0.92
EF	2.3	1.23
HC	2.3	0.48
PW	2.4	0.58
SS	2.2	0.69
Controls <	2.4	0.12

To be read in conjunction with Fig. 7.

binding assay (50). Comparison of the results from 256 diabetic sera obtained by radiobinding assay and by ELISA are shown in Fig. 8. Some individual sera showed wide differences but, in view of the fundamentally different assay kinetics discussed above, it is questionable whether one can expect a higher correlation between two such assays than that obtained here (0.74), and elsewhere (0.84) (54).

Comparison of Commercial and "In-house" Anti-Human Ig/Peroxidase Conjugates

The commercial conjugate used (Dako, Mercia Brocades, Weybridge, Surrey, U.K.) contained 15 mmolar sodium azide as preservative, whereas our "in-house" conjugates were stored as 1:2 concentrates in glycerol at -20°C without preservative. Azide inhibits ion exchange in metallo-enzymes (60).

HUMAN INSULIN

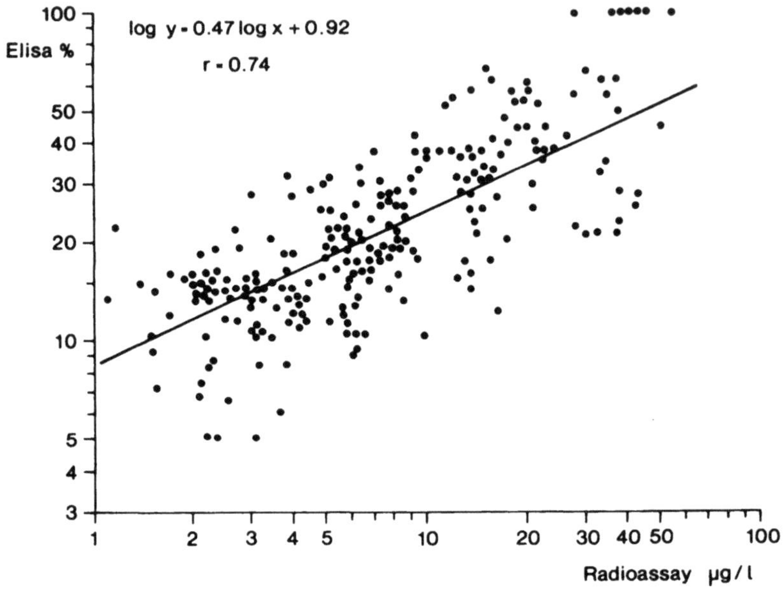

Fig. 8. Comparison of insulin-binding results from 256 sera obtained by radiobinding assay (RIA) and ELISA.

A direct comparison between the "in-house" and commercial conjugates, both at their optimum concentrations, is shown in Fig. 9. The assay using the commercial product appeared to suffer from a very high non-specific signal and consequently diminished dynamic range. In an attempt to confirm that azide at a concentration of 15 mmol/l was detrimental to the conjugate, our "in-house" product was diluted 1:2 in buffer containing 15 mmolar azide and stored overnight at 4°C before making it up to the operating dilution in azide-free buffer. Although the high non-specific binding seen with commercial conjugate was not apparent, there was a marked loss of signal strength

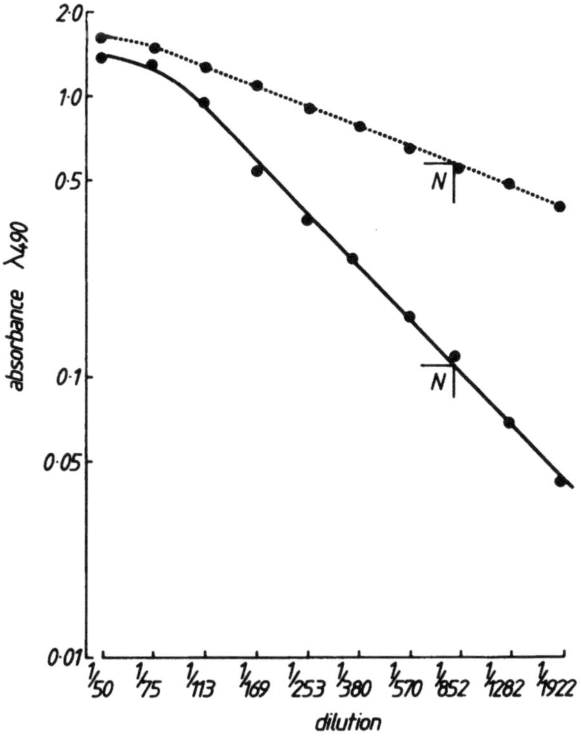

Fig. 9. Comparison of "in-house" (——) and commercial (Dako) (---) anti-human IgG conjugates. The commercial conjugate contains 15 mmolar sodium azide. The signals obtained from a single negative control serum (N) at a dilution of 1:20 are indicated.

(about 50% O.D.) in the assay (Fig. 10).

Modifications to ELISA

A number of modifications to this assay have been introduced and some of them validated.

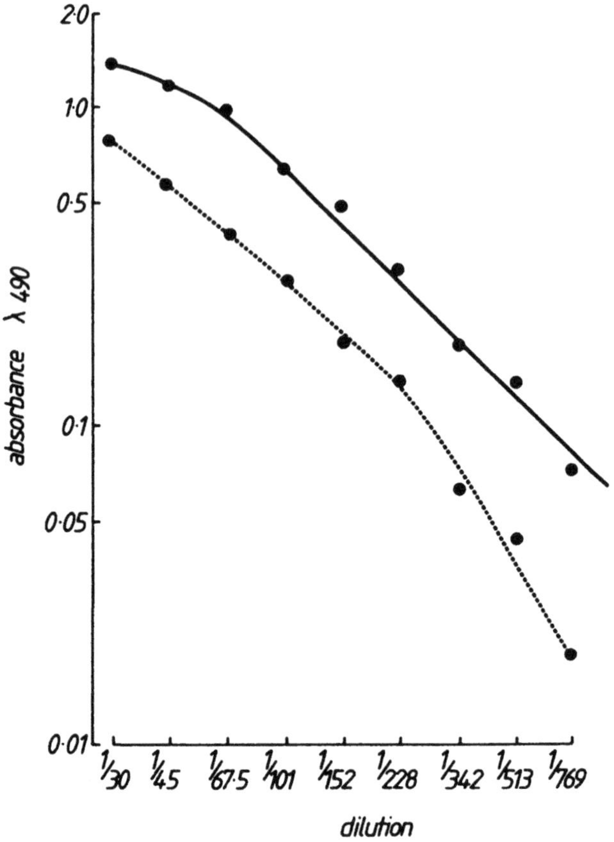

Fig. 10. The loss of signal strength resulting from dilution of "in-house" conjugate in buffer containing 15 mmolar sodium azide (---) as opposed to azide-free buffer (——).

Substrate: O-phenylene diamine (OPD) carries the theoretical risk of carcinogenicity and the practical disadvantage of requiring dimmed light during use. An alternative, tetramethyl benzidine (TMB) is non-carcinogenic and not light sensitive during use. It was made up from a light sensitive powder by adding 36 mg to 10 ml dimethyl sulfoxide (Analar quality).

This stock solution was stored in the dark and when required, diluted 1:100 in 0.1 M sodium acetate buffer, adjusted to pH 6.0 by addition of 0.1 M citric acid. To each 100 ml prepared solution were added 10 µl 30% H_2O_2. Fig. 11 shows a comparison of binding curves from a diabetic serum (temperature and incubation times as stated previously) using OPD and TMB, both at optimised concentrations. The reaction with OPD was maximal at 50 minutes whereas it was necessary to stop the reaction with TMB after 15 minutes. Temperature and incubation periods: a further modification was aimed at accommodating the assay within a day's work. Incubation periods were reduced and the temperature raised from 18°C to 37°C. By extending the substrate incubation period it was possible to obtain a satisfactory binding curve from serial dilutions of the standard serum commencing at 1:50.

Reference Serum Specificity and the Interpretation of Insulin Antibody Assays

Previous work in animals (73) and humans (11, 12) has shown how the insulin antibodies in different sera react with different epitopes on the insulin molecule. It has recently become clear to us, for example, that while antibodies induced by the administration to humans of heterologous insulins are directed primarily towards A chain determinants, autoantibodies to insulin are frequently specific for B chain determinants. Fig. 12 compares the binding curves of a diabetic serum and an insulin antoantibody serum to whole human insulin (panel a). Their similarity suggests that either would be suitable as a reference serum for the other in an ELISA for insulin antibody determination. Substitution of whole insulin by human B chain, however, shows that the autoantibody bound B chain residues strongly, while the diabetic serum bound only minimally (Fig. 12b); further substitution by desalinated insulin (B30 removed) restored binding by the diabetic serum but at the expense of bind-

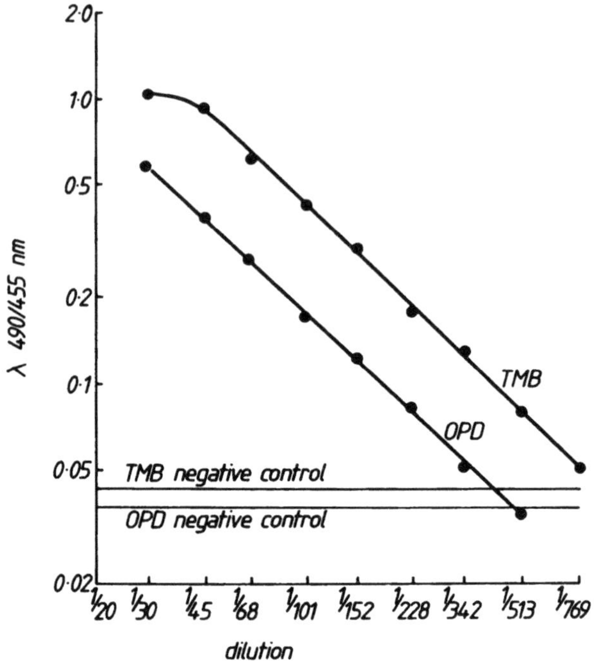

Fig. 11. The increase in signal strength and dynamic range obtained by using the peroxidase substrate tetramethyl benzidine (λ 490/455 nm) in place of O-phenylene diamine.

ing by the autoantibody. Thus the two sera appeared to bind different epitopes, making their use as mutual reference sera analytically meaningless. Finally, combination of the two sera, without change in their concentration, produced a signal considerably stronger than either gave alone. Clearly, broad specificity in a reference curve is desirable to ensure not only qualitative but also quantitative meaning to the result.

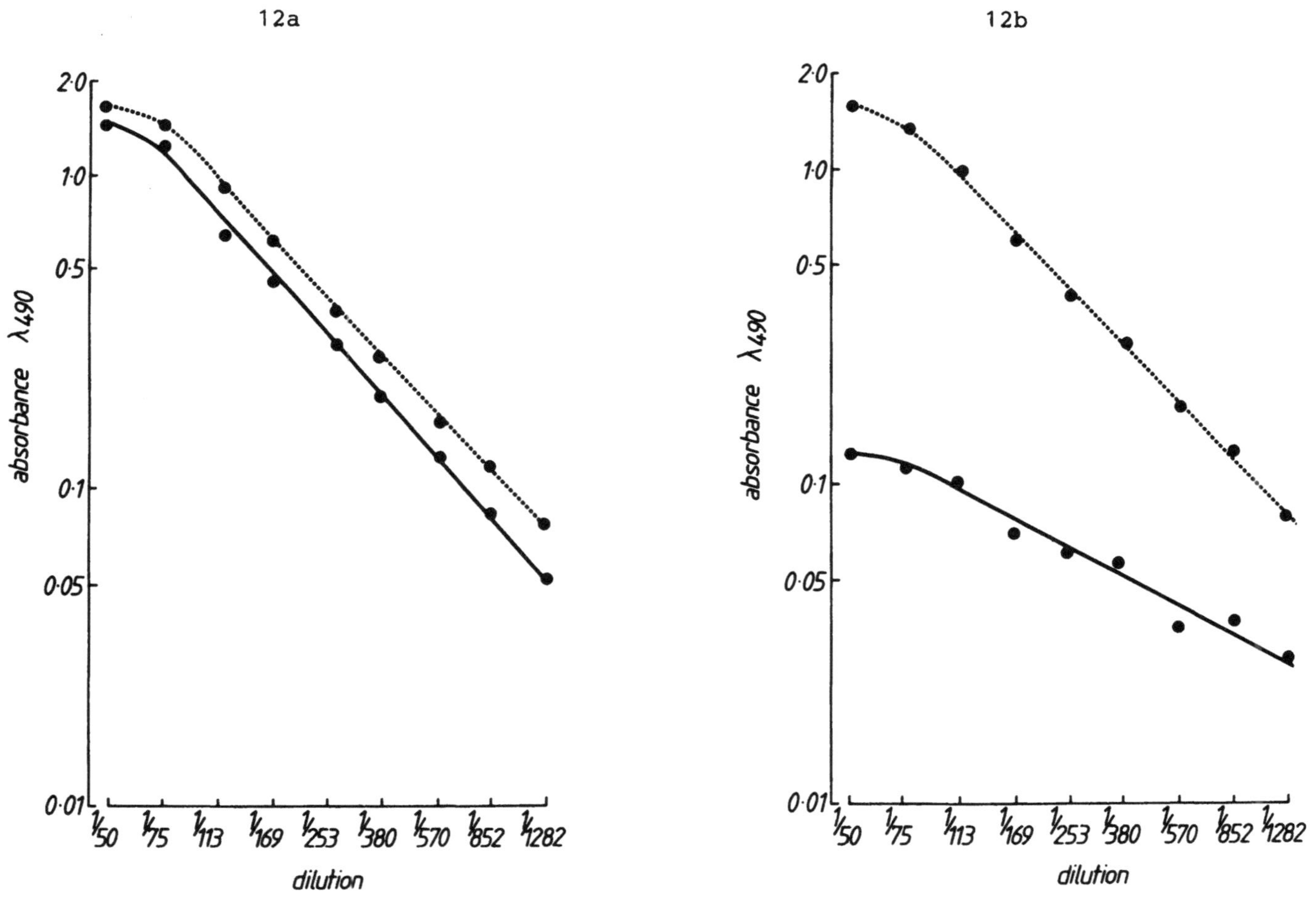

Fig. 12 (a). Comparable binding curves against whole human insulin obtained with mixed sera from five insulin-treated diabetics (——) and a subject with human-specific insulin autoantibodies (---).
(b). Binding curves obtained by incubating the same material as above with human insulin B chain. Binding by the diabetic sera (——) is much reduced, but by the autoantibody (---) unchanged.

12c 12d

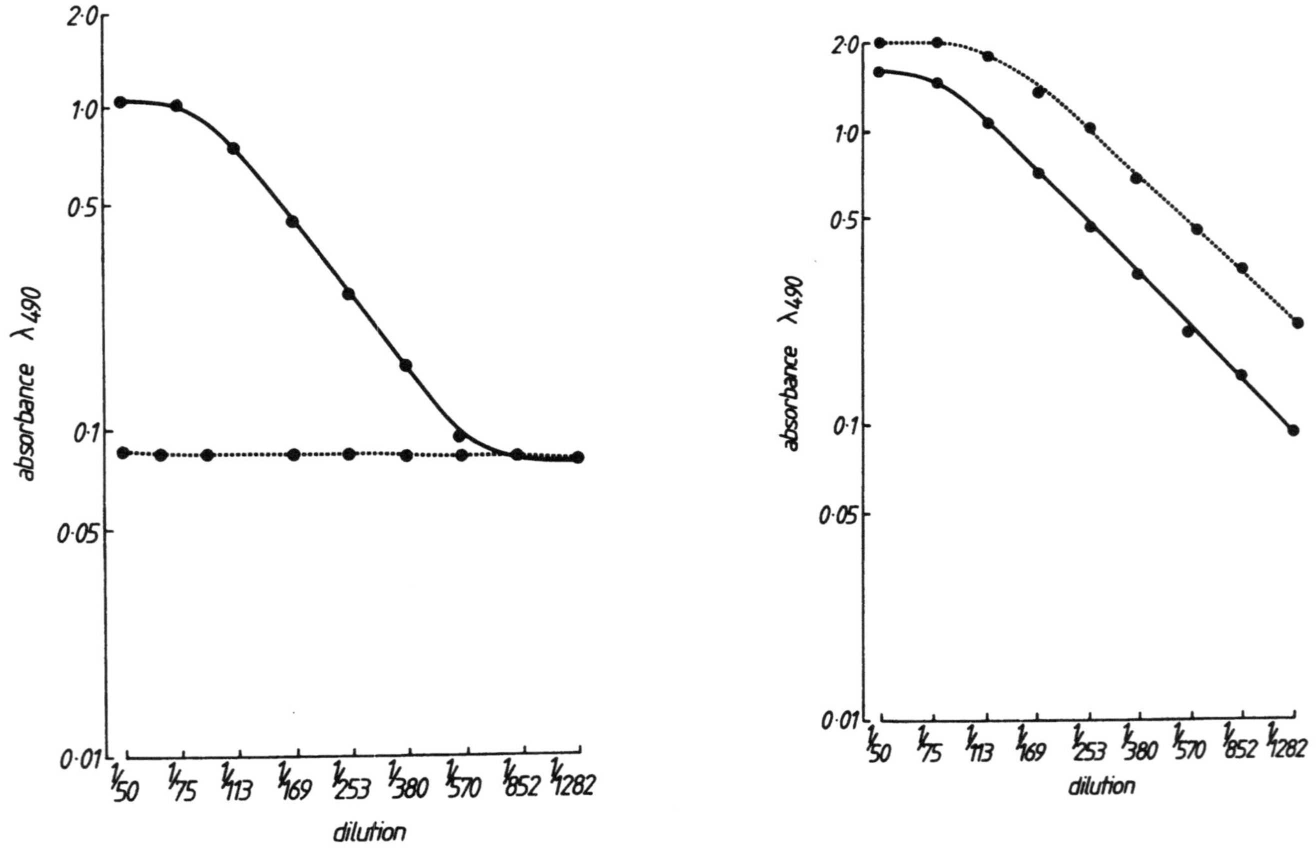

Fig. 12 (c). Binding curves obtained by incubating the same material as used in panel "a" with human insulin from which the THR-B30 residue had been removed. The autoimmune serum (---) no longer binds, confirming its specificity for THR-B30.

(d) Binding curves against whole human insulin obtained with the mixed sera from diabetics, before (——) and after (---) enrichment with the THR-B30 specific autoimmune serum.

General Comments and Conclusions

Insulin antibodies are often found in the sera of insulin-treated diabetics, although their frequency varies (28), depending among other factors on the purity of insulin administered (73) and the responsiveness of the individual (74). While low levels are of little clinical importance, high levels of antibody have been associated with insulin resistance and lipoatrophy (2, 4, 75). Insulin antibody studies are also providing insights into basic immunology. Insulin is a molecule of known tertiary structure and wide species variation. Many investigators have used it to demonstrate the fine specificity of immune responsiveness and its dependence on T cells (43), and macrophages (76).

Interest is now turning to autoantibodies to insulin in man (8, 9, 10). They show restricted clonality and appear to bind different determinants from those bound by antibodies induced during insulin treatment (12). Being spontaneous, autoantibodies may provide a more relevant insight into the human autoimmune process. Clinically, autoantibodies to insulin may have an important role as markers for islet-cell autoimmunity and diabetes (8, 9).

The interpretation of insulin antibody measurements demands considerable care. Non-competitive antibody excess systems have the advantage of simplicity. They measure directly the amount of ligand bound by a given volume of test serum. However, although the quality of a ligand can be standardised by insistence on purity, small variations in the quantity added can lead to significant shifts in the assay result. Antigen excess assays, by contrast, measure the amount of antibody bound by excess antigen. Expression of the result in terms of a negative serum, although theoretically attractive, may lead to loss of precision through day-to-day variation in non-specific binding. An alternative is the ratio between the signals from the test serum and a positive reference serum (ELISA index, 59). Test and reference sera, however, may bind to different determinants. As a consequence, the index derived for one test serum may prove artificially high or low in relation to that of another, depending on how closely the determinants bound by each match those bound by the reference serum. We have attempted to resolve this difficulty for ELISA by using a 'universal' reference serum containing specificities to a broad range of insulin determinants.

This chapter has reviewed currently available assays for insulin antibodies, with special reference to ELISA. The reasons for emphasizing ELISA are its low cost, simplicity and greater sensitivity. Absence of the administrative restrictions applied to isotope handling and ligand labelling is an extra advantage, and enzyme conjugates have a shelf-life many times that of radioiodine-labelled ligands. Apart from greater sensitivity, there are important theoretical considerations which favour ELISA for the measurement of insulin antibodies and autoantibodies. Chief among its disadvantages are the variations in coating protein adhesion and the necessity for repeated washing at each step. Washing can be automated with improved precision, and protein coating concentrations can be adjusted so as to make minimal difference to precision. Overall reproducibility is comparable to radioimmunoassay, provided that care is taken to optimise each step.

Acknowledgements

I wish to acknowledge the following for their financial support, advice, and technical assistance: The Wellcome Trust: Eli Lilly Company; Mr. Jose-Luis Diaz, Mrs. Wendy Couper, Mrs. Cecily Casey, Dr. Geoff Bull, Dr. Guy Dodson, Dr. Freda Stevenson and Dr. Gordon Reeves.

References

1. Berson, S.A., Yalow, R.S., Bauman, A., Rothschild, M.A., Newerly, K.: Insulin I^{131} metabolism in human subjects: demonstration of insulin-binding globulin in the circulation of insulin treated subjects. J. clin. Invest. 35, 170-190 (1956).
2. Berson, S.A., Yalow, R.S.: Quantitative aspects of the reaction between insulin and insulin-binding antibody. J. clin. Invest. 38, 1996-2016 (1959).
3. Dolovich, J., Schnatz, J.D., Reisman, R.E., Yagi, Y., Arbesman, C.E.: Insulin allergy and insulin resistance. J. Allergy 46, 127-137 (1970).
4. Reeves, W.G., Allen, B.R., Tattersall, R.B.: Insulin-induced lipoatrophy : evidence for an immune pathogenesis. Brit. med. J. 280, 1500-1503 (1980).
5. Kurtz, A.B., Matthews, J.A., Mustaffa, B.E., Daggett, P.R., Nabarro, J.D.N.: Decrease of antibodies to insulin, proinsulin and contaminating hormones with changing treatment from conventional beef to purified pork insulin. Diabetologia 18, 147-150 (1980).
6. Peacock, I., Taylor, A., Tattersall, R.B., Douglas, C.A., Reeves, W.G.: Effects of new insulins on insulin and C-peptide antibodies, insulin dose and diabetic control. Lancet 1, 149-152 (1983).
7. Schernthaner, G., Borkenstein, M., Fink, M., Mayer, W.R., Menzel, J., Schober, E.: Immunogenicity of human insulin (Novo) or pork monocomponent insulin in HLA-DR-typed insulin dependent diabetic individuals. Diabetes Care 6, suppl. 43-48 (1983).
8. Palmer, J.P., Asplin, C.M., Clemons, P., Lyen, K., Tatpati, O., Raghu, P.K., Paquette, T.L.: Insulin antibodies in insulin dependent diabetics before insulin treatment. Science 222, 1337-1339 (1983).
9. Wilkin, T.J., Hoskins, P.J., Armitage, M., Rodier, M., Casey, C., Diaz, J-L., Pyke, D.A., Leslie, R.D.G.: Value of insulin autoantibodies as serum markers for insulin-dependent diabetes mellitus. Lancet 1, 480-482 (1985).
10. Wilkin, T.J., Nicholson, S.: Autoantibodies against human insulin. Brit. med. J. 288, 349-352 (1984).
11. Kurtz, A.B., Matthews, J.A., Nabarro, J.D.N.: Insulin binding antibody : reaction differences with bovine and porcine insulins. Diabetologia 15, 19-22 (1978).
12. Diaz, J-L., Wilkin, T.J.: Species and epitope specificity of autoantibodies to insulin (in preparation).

13. Farr, R.S.: A quantitative immunochemical measure of primary interaction between 1*BSA and antibody. J. infect. Dis. 103, 239-242 (1958).

14. Dayhoff, M.O.: Atlas of protein sequence and structure. Vol. 5, Suppl. 3. National Biomedical Research Foundation, Washington (1972).

15. Reeves, W.G.: Immunology of diabetes and insulin therapy. In: "Recent Advances in Clinical Immunology", Ed. Thompson, R.A., Churchill Livingstone, London, 2nd edn., pp. 183-220 (1980).

16. Steiner, D.F.: Insulin today. Diabetes 26, 322-340 (1970).

17. Bloom, S.R., Adrian, T.E., Barnes, A.J., Polak, J.M.: Autoimmunity in diabetics induced by hormonal contaminants of insulin. Lancet 1, 14-17 (1979).

18. Schlichtkrull, J.: Antigenicity of monocomponent insulins. Lancet 2, 1260-1261 (1974).

19. Goeddel, D.V., Kleid, D.G., Bolivar, F., Heyneker, H.L., Yansura, D.G., Crea, A., Hirose, T., Kraszewski, A., Itakura, K., Riggs, A.D.: Expression in E. Coli of chemically synthesized genes for human insulin. Proc. natn. Acad. Sci. U.S.A. 76, 106-110 (1979).

20. Morihara, K., Oka, T., Tzuzuki, H.: Semisynthesis of human insulin by trypsin catalysed replacement of Ala-B30 by Thr in porcine insulin. Nature 280, 412-413 (1979).

21. Gattner, H.G., Danho, W., Naithani, V.K.: Enzyme-catalysed semi-sythesis with insulin derivatives. In: "Chemistry, Structure and Function of Insulin and Related Hormones", Proc. Second International Insulin Symposium, Aachen, West Germany, Ed. Brandenburg, D., Wollner, A., Walter de Gruyter, Berlin, New York, pp. 117-123 (1980).

22. Hagedorn, H.C., Jensen, B.N., Krarup, N.B., Wodstrup, I.: Protamine insulinate. J. Am. med. Ass. 106, 177-180 (1936).

23. Hallas-Moller, K., Petersen, K., Schlichkrull, J.: Crystalline and amorphous insulin-zinc compounds with prolonged action. Science 116, 394-398 (1952).

24. Lockwood, D.H., Prout, T.E.: Antigenicity of heterologous and homologous insulin. Metabolism 14, 530-538 (1965).

25. Deckert, T., Grundahl, E.: The antigenicity of pig insulin. Diabetologia 6, 15-20 (1970).

26. Steiner, D.F., Hallund, O., Rubinstein, A., Cho, S., Bayliss, C.: Isolation and properties of proinsulin, intermediate forms and other minor components from crystalline bovine insulin. Diabetes 17, 725-731 (1968).

27. Kurtz, A.B., Nabarro, J.D.N.: Circulating insulin-binding antibodies. Diabetologia 19, 329-334 (1980).

28. Andersen, O.O.: Insulin antibody formation. 1. The influence of age, sex, infections, insulin dosage and relation to diabetes. Acta endocr. 71, 126-140 (1972).

29. Menzel, R., Knopse, S., Ziegler, M., Wilke, W., Michael, R.: Failure of appearance of insulin antibodies in dogs adapted to bovine-porcine insulin. Diabetologia 7, 386-390 (1971).

30. Goldman, J., Carpenter, F.H.: Zinc binding circular dichroism and equilibrium sedimentation studies on insulin (bovine) and several of its derivatives. Biochemistry 13, 4566-4574 (1974).

31. Adams, M.J., Blundell, E.J., Dodson, E.J., Dodson, G.G., Vyayan, M., Baker, E.N., Harding, M.M., Hodgkin, D.C., Rimmer, B., Sheat, S.: Structure of rhombohedral 2 zinc insulin crystals. Nature 224, 491-495 (1967).

32. Huget, R., Opitz, H.G., Flad, H.D.: Adjuvant and suppressor activity of the polycation protamine hydrochloride in the primary immune response of mice. Zeitschrift für Immunitatforschung Experimentelle und Klinische Immunologie 152, 190-199 (1976).

33. Arquilla, E.R., Thiene, A., Brugman, T., Ruess, W., Sugiyama, R.: Effects of zinc ion on the conformation of antigenic determinants on insulin. Biochem. J. 175, 289-297 (1978).

34. Hirata, Y., Ishizu, H.: Elevated insulin-binding capacity of serum proteins in a case of spontaneous hypoglycaemia and mild diabetes not treated with insulin. Tohoku J. Exp. Med. 107, 277-286 (1972).

35. Keck, K.: Ir-gene control of immunogenicity of insulin and A-chain loop as a carrier determinant. Nature 254, 78-79 (1975).

36. Roitt, I.M.: In: "Essential Immunology", Blackwell Scientific Publications, Oxford, 4th edition, pp. 73-74 (1980).

37. Schroer, J.A.: Hybridoma antibody recognition of the insulin molecule. In: "Monoclonal Antibodies in Endocrine Research", Eds. Fellows, R., Eisenbarth, G., Raven Press, New York, pp. 167-179 (1981).

38. Nye, L., De Carvalho, L.P., Roitt, I.M.: An investigation of the clonality of human autoimmune thyroglobulin antibodies and their light chains. Clin. Exp. Immunol. 46, 161-170 (1981).

39. Faulk, W.P., Karam, J.H., Fudenberg, H.H.: Human anti-insulin antibodies. J. Immunol. 106, 1112-1126 (1971).

40. Irvine, W.J., Di Mario, U., Feek, C.M., Ting, A., Morris, P.J., Gray, R.S., Duncan, L.J.P.: Insulin antibodies in relation to islet antibodies and HLA antigens in insulin-dependent (type I) diabetes. J. Clin. Lab. Immunol. 1, 111-114 (1978).

41. Schernthaner, G., Ludwig, H., Mayer, W.R.: Immunoglobulin-G antibodies and immune region - associated alloantigens in insulin dependent diabetes mellitus. J. clin. Endocrinol. Metab. 48, 403-407 (1979).

42. Kahn, C.R., Mann, D., Rosenthal, A.S., Galloway, J.A., Johnson, A.H., Mendell, N.: The immune response to insulin in man. Diabetes 31, 716-723 (1982).

43. Jensen, P.E., Kapp, J.A.: Regulatory mechanisms of the immune response to heterologous insulins. Cell. Immunol. 87, 73-84 (1984).

44. Asplin, C.M., Dornan, T.L., Raghu, P.K., Hansen, J.A., Palmer, J.P.: The antibody response to insulin therapy: a prospective study in HLA-typed insulin dependent diabetic subjects. Diabetes 33, 966-969 (1984).

45. Rosenthal, A.S., Barcinski, M.A., Rosenwasser, L.: Function of macrophages in genetic control of immune responsiveness. Fed. Proc. 37, 79-85 (1976).

46. Kilpatrick, J.M., Virella, G.: Isolation and characterisation of soluble insulin-anti-insulin immune complexes formed in vitro and in vivo in sera from patients with diabetes mellitus. Clin. Exp. Immunol. 40, 445-452 (1980).

47. Irvine, W.J., Al-Khaleeb, S.F., Di Mario, U., Feek, C.M., Gray, R.S., Edmond, B., Duncan, L.J.P.: Soluble immune complexes in the sera of newly diagnosed insulin-dependent diabetics and in treated diabetics. Clin. Exp. Immunol. 30, 16-21 (1977).

48. Folling, I.: Insulin-anti-insulin complexes. Acta endocr. 83, (Suppl. 205), 199-209 (1976).

49. Harwood, R.: Insulin-binding antibodies and "spontaneous" hypoglycaemia. New Engl. J. Med. 262, 978-979 (1960).

50. Reeves, W.G., Kelly, U.: An immunochemical method for the quantitation of insulin antibodies. J. Immunol. Methods 34, 329-338 (1980).

51. Maron, R., Elias, D., de Jongh, B.M., Bruining, G.J., van Rood, J.J., Schechter, Y., Cohen, I.R.: Autoantibodies to the insulin receptor in juvenile onset insulin-dependent diabetes mellitus. Nature (London) 303, 817-818 (1983).

52. Plotz, P.H.: Autoantibodies are anti-idiotype antibodies to antiviral antibodies. Lancet 2, 824-826 (1983).

53. Reeves, W.G.: Insulin antibody determination, theoretical and practical considerations. Diabetologia 24, 399-403 (1983).

54. Flynn, S.D., Keven, D.F., Torretti, B., Dieterle, R.C., Grauds, S.: Factors affecting enzyme-linked immunosorbent assay for insulin antibodies in human plasma. Clin. Chem. 27, 1753-1757 (1981).

55. Engrall, E., Jansson, K., Perlmann, P.: Enzyme-linked immunosorbent assay. II: Quantitative assay of protein antigen, immunoglobulin G, by means of enzyme-labelled antigen and antibody coated tubes. Biochim. biophys. Acta 251, 427-434 (1971).

56. Baxter, R.C., Yue, D.K., Turtle, J.R.: Equilibrium binding studies of insulin antibodies in diabetes subjects. Clin. Chem. 22, 1089-1094 (1976).

57. Bullock, S.L., Walls, K.W.: Evaluation of some of the parameters of the enzyme-linked immunosorbent assay. J. infect. Dis. 136 (S), 279-285 (1977).

58. De Savigny, D., Voller, A.: The communication of ELISA data from laboratory to clinician. J. Immunoassay 1, 105-128 (1980).

59. McLachlan, S.M., Clark, S., Stimson, W.H., Clark, F., Smith, B.R.: Studies of thyroglobulin autoantibody synthesis using a micro-ELISA assay. Immunol. Lett. 4, 27-33 (1982).

60. Skom, J.H., Talmage, D.W.: Nonprecipitating insulin antibodies. J. clin. Invest. 37, 783-786 (1958).

61. Christiansen, Aa.H.: Radioimmunoelectrophoresis in the determination of insulin-binding to IgG. Methodological studies. Horm. Metab. Res. 5, 147-154 (1973).

62. Ooms, H.S., Arquilla, E.R.: Labelled proteins in tracer studies. Atomic Energy Community, Euratom, p. 237 (1966).

63. Brunfeldt, K., Hansen, B.A., Jorgensen, K.R.: The immunological reactivity and biological activity of iodinated insulin. Acta endocr. 57, 307-329 (1968).

64. Andersen, O.O., Brunfeldt, K., Abildgard, F.: A method for quantitative determination of insulin antibodies in human plasma. Acta endocr. 69, 195-208 (1972).

65. Nakagawa, S., Saito, N., Nakayama, H., Sasaki, T., Watanabe, T., Aoki, S.: Detection of IgE insulin antibody with radioallergosorbent test. Diabetologia 14, 33-38 (1978).

66. Falholt, K.: Determination of insulin-specific IgE in serum of diabetic patients by solid phase radio-immunoassay. Diabetologia 22, 254-257 (1982).

67. Hudson, L., Hay, F.C.: Practical Immunology. Blackwell Scientific Publications, Oxford, 2nd edition, pp. 237 (1980).

68. Salacinski, P.R., McLean, C., Sykes, J.E., Clement-Jones, N., Lowry, P.J.: Iodination of proteins, glycoproteins and peptides using a solid-phase oxidising agent 1, 3, 4, 6-tetrachloro 3α, 6α-diphenyl glycoluril (Iodogen). Annal. Biochem. 117, 136-146 (1981).

69. Chessum, B.S., Denmark, J.R.: Inconstant ELISA. Lancet i, 161 (1978).

70. Kricka, L.J., Carter, T.J.N., Burt, S.M., Kennedy, J.H., Holder, R.H., Halliday, M.I., Telford, M.E., Wisdom, G.B.: Variability in the absorption properties of microtitre plates used as solid supports in enzyme immunoassay. Clin. Chem. 26, 741-744 (1980).

71. Dixon, K.: Measurement of antibodies to insulin in serum. Clin. Chem. 20, 1275-1281 (1974).

72. White, A., Handler, P., Smith, E.L.: In: "Principles of Biochemistry", McGraw-Hill, New York, 4th edition, pp. 236-240 (1968).

73. Yue, D.K., Turtle, J.R.: New forms of insulin, and thèir use in the treatment of diabetes. Diabetes 26, 341-345 (1977).

74. Reeves, W.G., Gelsthorpe, K., Van der Minne, P., Torensma, R., Tattersall, R.B.: HLA phenotype and insulin antibody production. Clin. Exp. Immunol. 57, 443-448 (1984).

75. Witters, L.A., Ohman, J.L., Wir, G.C., Raymond, L.W., Lowell, F.C.: Insulin antibodies in the pathogenesis of insulin allergy and resistance. Am. J. Med. 63, 703-709 (1977).

76. Rosenthal, A.S.: Determinant selection and macrophage function in genetic control of the immune response. Immunol. Rev. 40, 136-152 (1978).

IMMUNOSELECTIVE ELECTRODES

D. Monroe
University of Tennessee Center for the Health Sciences, Department of Medicine, Infectious Diseases and Connective Tissues Sections, Memphis, Tennessee, U.S.A.

Many analytical methods have been developed for accurately measuring trace quantities of protein in biological fluids. Bioselective electrochemical probes are among the newer sensing devices that have enabled scientists to rapidly and precisely determine, not only specific ions or dissolved gases, but also proteins and immunological reactions (1, 2). Recent advances in protein chemistry, enzymology, and immunochemistry have also led to production of very sensitive enzyme immunoassay (EIA) methods for measuring minute amounts of antibodies, antigens, and haptens (3, 4). Marriage of bioselective electrode technology with immunoassay methods has culminated in instruments that are easily prepared, highly selective, specific, and sensitive. These novel analytical immunoselective probes have much to offer and are proving to be excellent practical methods of analysis with many applications.

Electrode Origin

The idea for immunoselective sensors originated in experiments where electrical potential changes were first noted when specific antigens became bound to sensitized metal electrodes (23). Such affinity immunoelectrodes are easily made by covalently binding biospecific ligands to a titanium (5) or platinum (7) wire. Some of these earlier electrodes were used for conducting

Immunoassay Technology, Vol. 2

immunoassays on serum albumin (5, 6), creatine kinase MB (7), anti-benzoate antibodies (8), Wasserman antibodies (9), and anti-DNP antibodies (10, 11). Others included immunoassays by differential pulse polarography (12) and antibody binding measurements with hapten-selective membrane electrodes (13).

Greater sensitivity of potentiometric immunoassays became a reality when enzyme catalysts were used in the analysis (Table 1). These enzyme immunoselective electrodes combine the sensitivity and specificity of EIA with the advantages of electrochemical sensor techniques that monitor enzymatic reaction end products. Ion or gas selective electrodes usually comprise the internal components of these systems.

Table 1. Substrate detection limits of electrode enzyme labels.

Enzyme label	Substrate	Detection limits (mole/l)	Electrode sensor	Reference
Asparaginase	Asparagine	3.0×10^{-14}	Ammonia	19
Urease	Urea	3.5×10^{-14}	Ammonia	19
Adenosine Deaminase	Adenosine	1.6×10^{-13}	Ammonia	19
Acetylcholin-Esterase	Acetylcholine	1.0×10^{-8}	pH	20
G-6-P Dehydrogenase	G-6-P/NAD^+	1.0×10^{-5}	Pt/Ag/AgCl	17
HRP	H_2O_2	1.0×10^{-11}	I_2	22

An electrode EIA reported in 1979 made use of horseradish peroxidase (HRP) antibody conjugates and an iodide-sensitive electrode for detecting hepatitis B surface antigen (14, 15). Such antigens in biological fluids were easily measured when using an EIA sandwich molecule procedure. Enzymatic activity could be detected by the electrode after potassium iodide and hydrogen peroxide substrate were added. This new approach in EIA proved more sensitive, practical,and less time-consuming than most immunoassays. Even turbidity or nonspecifically interfering factors could easily be eliminated by this method.

The simplicity and sensitivity of EIAs was shown in the early 1980s to be largely determined by the kind of enzyme label employed (16). EIAs are often limited by those enzymes that act on chromogenic substrates with end products having a colour change detectable only by spectrophotometric means. Electrode-based EIAs, however, are more versatile since many different labels may be included. Dehydrogenase (17), lysozyme (18), and deaminating enzyme labels (19) for potentiometric EIAs have helped to increase test stability, activity, and binding ability.

Affinity Sensor

Affinity immunoselective sensors are simple in construction. A glass pH electrode or metal wire provides the internal electrical component for these sensing devices. Purified proteins, monotypic antibodies, or isolated antigens can be adsorbed or covalently bound by simple soaking procedures (1, 2). Proteins also bound to silicon chips (24) and/or nitrocellulose membranes (26) are often employed by fixation to the electrode surface. Antigen antibody affinity and immunochemical interactions occurring at the probe's tip will be converted into an electrical impulse for analysis.

An antibody electrode can easily "fish" out type specific antigens from a solution mixture (Fig. 1). A quantitative antigen level is electronically determined by the amount of antigen being caught with electrode fixed antibody molecules. Purified antigen electrodes can be built for similarly measuring specific antibodies in a mixture. Such affinity electrodes are quite stable and reusable by dissociation of the immune-complex formed at the electrode surface. This is easily accomplished after each measurement by simply immersing the spent electrode into a solution of low pH, urea, or high concentration of chaotropic

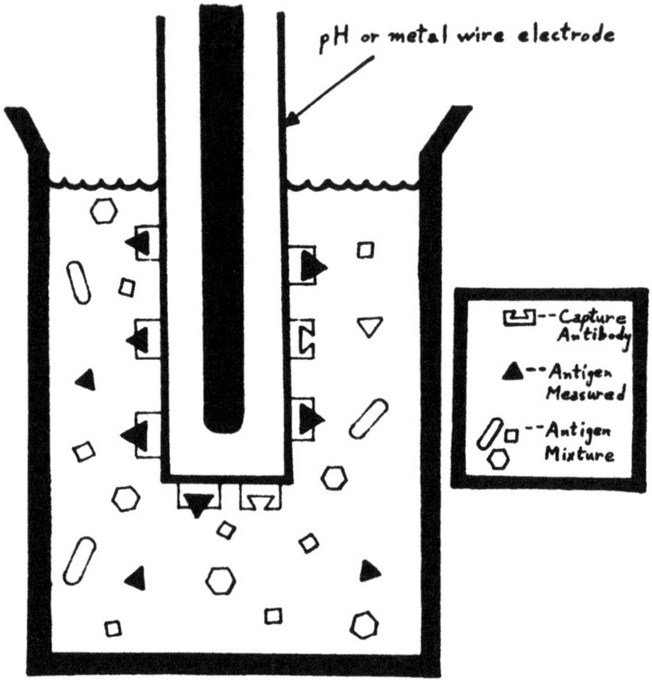

Fig. 1. Antibody affinity immunoselective electrode.

ions (27). Antigen antibody dissociation is very rapid and the intact electrode is ready for reuse in a matter of minutes.

Steady state electrical potentials are usually reached after 5 to 10 minutes of immersion into a test solution. Response times are reproducible and detection limits within the microgram/ml range (1, 2). Temperature, pH, and ionic strength affect immune-complex formation and should be carefully controlled for obtaining maximum electrode response.

EIA Sensor

Antibodies, antigens, and haptens can be precisely quantitated by the EIA technique using an enzyme label and either a pH, ion, or gas selective electrode to follow enzymatic activity. Potentiometric enzyme monitoring offers greater measurement accuracy than affinity sensors due to the sensitivity amplification effect of an enzyme catalyst. Potentiometric EIA detection limits are determined in microgram/litre values while affinity immunosensors and spectrophotometric EIAs are measured within the microgram/millilitre to nanogram/millilitre range. A good understanding of spectrophotometric EIA methodology is often necessary, not only for developing the proper potentiometric EIA system required, but also for appreciating the many advantages which electrode-based EIAs have to offer. Shorter assay times, greater sensitivity, wider variety of enzyme label use available, kinetic studies made possible, and nonspecific interfering factors eliminated are but a few benefits enjoyed by those investigators performing potentiometric EIAs (Table 2).

EIA sensors use a competitive or double antibody "sandwich" technique for detection and quantitation of antigens or antibodies. When a known amount of enzyme labelled antigen is mixed with an unlabelled antigen test mixture, competitive binding of both antigen types for the electrode fixed antibody takes place (Fig. 2). An enzyme labelled bound immune-complex will form on the probe's tip, usually within less than 5 to 10 minutes. The electrode surface is then washed several times with distilled water or buffer before being placed into a specific substrate solution.

The kind of substrate employed is largely determined by the enzyme label used, as well as by the available type of internal sensing electrode component capable of recording the enzyme-substrate end-products obtained. Enzyme digestion of the substrate occurs when the probe is placed in the substrate solution. As a result, gases may be evolved, pH changes may take place, or an ionic shift may be detected. Variable recorded

Table 2. EIA Detection system comparison

Potentiometric	Spectrophotometric
Detection limits are often in microgram/litre or less range.	Microgram/ml to nanogram/ml levels usually measured.
Short assay time (5-10 min) after substrate added. Reproducible starting potentials achieved 2-3 min between assays. Electrode response time major rate limiting factor.	Enzyme activity measured over longer time period after substrate added (30 min to 2 hr) separate assays read rapidly, ≤30 sec per test.
Allows true kinetic assay studies since initial reaction rates easily monitored and followed to completion.	Enzyme activity limited to fixed-time studies following substrate incubation. Kinetic studies are difficult.
Nonspecific interfering factors are eliminated. Turbidity is not a problem when using electrode methods.	Turbid solid phase support system and immune-complex formation may interfere with readings.
Allows for a variety of enzyme labels to be used. Enzyme needed is determined by the substrate and electrochemical sensor used.	Enzyme label limited to only those acting on chromogenic substrates with a resulting detectable colour change.
Electrochemical measurements performed directly in enzyme solutions (0.8-1.0 ml vols.) without colorimetric reagents added or drastic changes (pH, etc.) in solution condition.	Light measurements made through reaction solutions (0.1-0.25 ml) often changed when colorimetric reagents are added.
Excellent analytical precision, sensitivity, specificity, and	Test precision, sensitivity and specificity excellent.

Table 2. EIA Detection system comparison (continued)

Potentiometric	Spectrophotometric
reproducibility. Day-to-day variability often ≤5% with relative standard deviation ≤10% for previous potentiometric activity.	Reproducibility good but polystyrene quality dependent and may vary with cuvette type used.

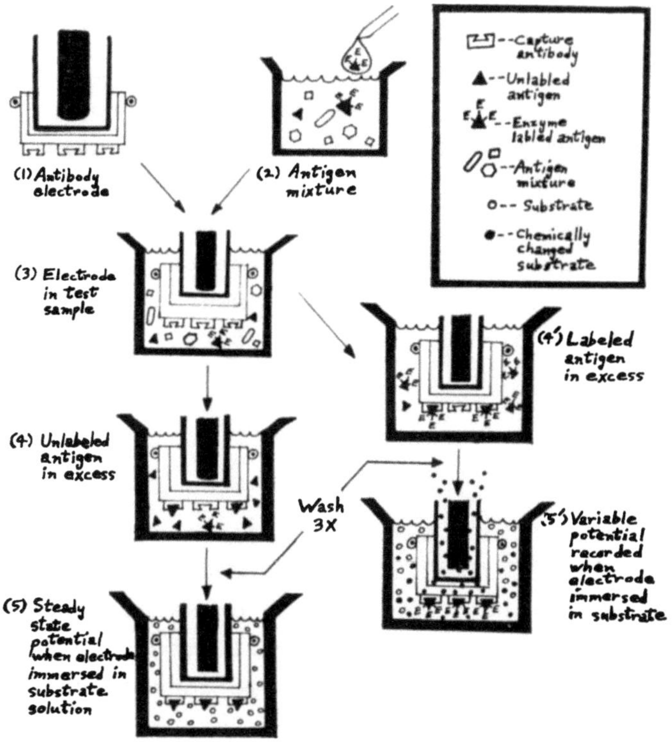

Fig. 2. Competitive inhibition EIA electrode.

potentials occur until a steady state reaction is completed, which is approximately 10 minutes or less. Reaction rates can be continuously followed and, consequently, kinetic studies easily made possible by this technique.

Unknown antibody amounts can likewise be determined in a similar manner or by means of an often more sensitive "sandwich" competitive method. In the latter case, known unlabelled antigen quantities are first bound to the electrode fixed antibody (species different from that being measured) to form an immune-complex at the probe's tip. Precise amounts of enzyme labelled antibodies introduced into the antibody test mixture equally compete for binding with the probe immune-complex. An enzyme labelled double antibody sandwich with the antigen located in-between is then formed. After several washings with distilled water or buffer, the probe is immersed into an appropriate substrate. Resulting potentials are used for quantitating the unknown amount of antibody present in the test mixture. Competitive, sandwich, direct, indirect, or a variety of EIA methods together with potential measurements can be employed by the investigator to obtain the most sensitive, specific, and inexpensive potential EIA system desired.

Applications

Potentiometric competitive EIAs have been employed primarily in the area of clinical medical research. They have been used for monitoring serum samples from patients on maintenance drug therapy (17), evaluating normal or abnormal pregnancy and discovering tumours by measuring human chorionic gonadotropin hormone levels in biological samples (20), and determining the biologically important nucleotide, adenosine 3', 5'-cyclic-monophosphoric acid, cAMP (16). Other applications include estimating oestradiol-17β concentrations in different solutions (22), detecting hepatitis B surface antigen (HBsAg) in biological fluids (15), measuring immunoglobulin G (IgG) levels in human serum (21), and quantitating the clinically important steroid hormone cortisol (19). Electrode-based EIAs have also been used to detect a model hapten group, dinitrophenyl (DNP) (19) as well as a model protein antigen, bovine serum albumin (BSA) (16).

Carver Genetic Physics, a Seattle Washington (USA) based company, has developed an affinity monoclonal antibody conjugated electronic probe to be used as a diagnostic tool for the early detection of cancer (24). Antibodies attached to a silicon chip at the probe's tip bind tumour associated antigens from patient's blood serum. Disposable interchangeable tips coated with the appropriate monoclonal antibody allows for a wide range of tests.

These immunoselective probes can be used in a physician's office, clinical and research laboratories, or out-patient clinic. Most systems are portable, inexpensive, do not require radioisotopes, and probe tips are disposable or easily regenerated. Research scientists can affinity-purify antigens by coating tips with type specific monoclonal antibodies. When the tip becomes antigen saturated, a maximum steady-state electronical digital readout is obtained on the probe. Either antigens or antibodies can be purified in this manner, depending on the type of probe tip used.

Although the primary market for these probes includes medical research and diagnostics, they are also being developed to test for a wide range of contaminants, additives, toxins, and pollutants. They will also be used for quality control maintenance on foodstuffs, canned goods and agricultural products. Other applications include evaluating human thyroid function, measuring blood cortisol levels, and determining the severity of a heart attack by measuring the level of cardiac enzyme. As a cardiac holographic imager, the probe can be used to electronically create a three-dimensional sound field to simulate heart action, opening or closing of particular heart valves, and blood flow through the great arteries.

Automation

Automated immunoselective electrode systems using continuous flow analysis are a new concept that will soon become available for specific commercial use. They provide an effective means of rapidly and accurately quantitating large numbers of samples while maintaining low processing cost. Unit design, however, must be carefully considered since analysis output rate is greatly affected by variables such as flow rates, diluent reagents, sample viscosity, tubing length, dead space volumes and flow through areas. Electrode life expectancy, response time, and regeneration rates are additional critical factors that may hinder the obtaining of prompt test results. Present automated electrode EIA continuous flow systems include measuring human serum immunoglobulin (IgG) levels (21), monitoring phenytoin drug amounts in patients' samples (17), quantitating human hormone and steroid values (22), and making direct measurements of amino acids in plasma samples (25).

Potentiometric measurements are registered on a strip chart recorder taken on a pH/mV meter equipped with immunoselective indicator and reference electrodes that are enclosed in a continuous flow cell as described (22). Key components of this automated system (Fig. 3) consist of a multi-channelled proportioning pump, temperature controlled mixing coils, on-line Teflon dialysis unit, and the flow through electrode cell. Test samples, buffers, wash, enzyme conjugates, and electrode regeneration reagents are pumped through the continuous flow system in accordance with an automatic pre-programmed time sequence.

Samples are well mixed in appropriate dilution ratios with buffers and enzyme conjugates as they are being pumped through tubing coils into the automated system. Upon passage within the Teflon dialysis unit any dissolved gases and air will be eliminated by moving through the gas-permeable walls. Background ammonia, for example, will diffuse out and be converted into ammonium ions by an acid reservoir trap (25). The

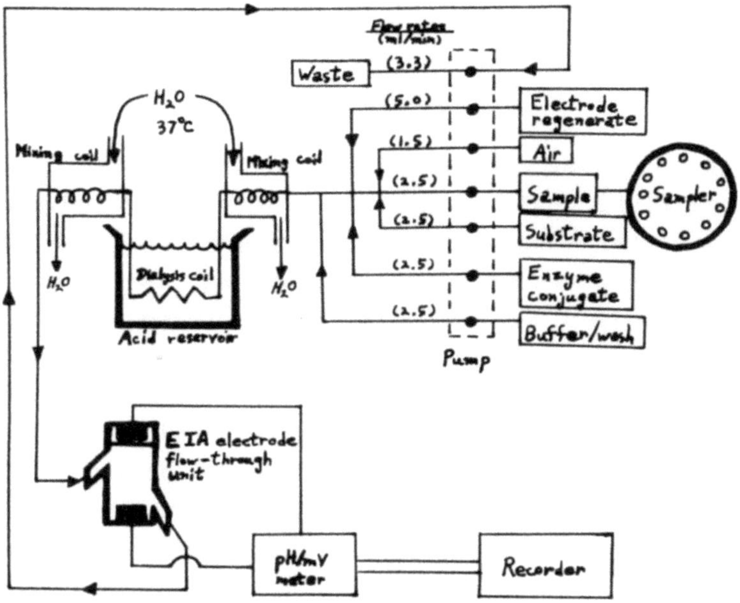

Fig. 3. Automated potentiometric EIA system.

gas free sample and enzyme conjugate compete for binding sites to antibody molecules fixed on the indicator immunoselective electrode surface. Any bound enzyme conjugate converts substrate being pumped into the electrode cell into gas end products easily detected by the internal gas sensing probe described earlier (Fig. 2). After a steady-state potential is reached and recorded, an electrode regenerating solution (acid, urea, or chaotropic ions) is then pumped into the system followed by a short purging wash cycle. The bound sample and/or enzyme conjugate are then removed from fixed electrode antibodies to waste and the spent electrode renewed for additional testing. This entire process takes place within a matter of minutes, with between 12 (21) to 30 (25) samples/hour being measured, depending upon the characteristics of the automated system used.

Future Outlook

Immunoselective probes are excellent practical, simple, low cost new analytical tools which will probably require greater interest and better understanding before gaining large scale acceptance. Further research, refinement, and use of specific monoclonal antibodies should definitely increase electrode immunoselective specificity and range. Extensively used in medical research for patient diagnosis and making macromolecular investigations, additional applications will include screening of environmental carcinogens, mutagens, industrial and agricultural pollutants.

Using automated continuous flow electrode based immunoassays will permit easy construction of small durable flexible catheters which can be implanted as biomedical transducers for monitoring patient therapy. Such microelectrode probes will be used for measuring very small volumes of samples without disturbing the sample environment. Only greater publicity, interest and further use will prove these probes to be priceless analytical tools.

References

1. Monroe, D.: Bioselective electrodes, part one. American Clinical Products Review 4, 24-34 (1985).
2. Monroe, D.: Bioselective electrodes, part two. American Clinical Products Review 4, 46-57 (1985).
3. Monroe, D.: ELISA, a versatile chemical tool. American Clinical Products Review 2, 22-27 (1983).
4. Monroe, D.: Enzyme immunoassay. Anal. Chem. 56, 920A-931A (1984).
5. Lowe, C.R.: Affinity electrode. FEBS letters 106, 405-408 (1979).

6. Alexander, P.W., Rechnitz, G.A.: Ion-electrode based immunoassay and antibody-antigen precipitin reaction monitoring. Anal. Chem. 46, 1253-1257 (1974).

7. Yuan, C.L., Kuan, S.S., Guilbault, G.G.: Immunoelectrochemical assay for creatine kinase isoenzyme MB. Anal. Chem. 53, 190-193 (1981).

8. Solsky, R.L., Rechnitz, G.A.: Preparation of antibody-selective electrode. Anal. chim. Acta 99, 241-246 (1978).

9. Aizawa, M., Suzuki, S.: An immunosensor for syphilis. J. Solid-Phase Biochem. 4, 25-31 (1979).

10. Solsky, R.L., Rechnitz, G.A.: Preparation and properties of an antibody-selective membrane electrode. Anal. chim. Acta 123, 135-141 (1981).

11. Solsky, R.L., Rechnitz, G.A.: Antibody-selective membrane electrodes. Science 204, 1308-1309 (1979).

12. Heineman, W.R., Anderson, C.W., Halsall, H.B.: Immunoassay by differential pulse polarography. Science 204, 865-866 (1979).

13. Meyerhoff, M., Rechnitz, G.A.: Antibody binding measurements with hapten-selective membrane electrodes. Science 195, 494-495 (1977).

14. Boitieux, J.L., Desmet, G., Thomas, D.: Potentiometric determination of hepatitis B surface antigen in biological fluids. Clin. chim. Acta 88, 329-336 (1978).

15. Boitieux, J.L., Desmet, G., Thomas, D.: An antibody electrode, preliminary report on a new approach in enzyme immunoassay. Clin. Chem. 25, 318-321 (1979).

16. Meyerhoff, M.E., Rechnitz, G.A.: Electrode-based enzyme immunoassay using urease conjugates. Meth. in Enzymology 70, 439-454 (1980).

17. Eggers, H.M., Halsall, H.B., Heineman, W.R.: Enzyme immunoassay with flow-amperometric detection of NADH. Clin. Chem. 28, 1848-1851 (1982).

18. Gebauer, C.R., Rechnitz, G.A.: Ion selective electrode estimation of avidin and biotin using a lysozyme label. Anal. Biochem. 103, 280-284 (1980).

19. Gebauer, C.R., Rechnitz, G.A.: Deaminating enzyme labels for potentiometric enzyme immunoassay. Anal. Biochem. 124, 338-348 (1982).

20. Mascini, M., Zolesi, F., Palleschi, G.: pH Electrode-based enzyme immunoassay for the determination of human chorionic gonadotropin. Anal. letters 15, 101-113 (1982).

21. Alexander, P.W., Maltra, C.: Enzyme-linked immunoassay of human immunoglobulin G with the fluoride ion selective electrode. Anal. Chem. 54, 68-71 (1982).

22. Boitieux, J.L., Lemay, C., Desmet, G., Thomas, D.: Use of solid phase biochemistry for potentiometric enzyme immunoassay of estradiol-17β - preliminary report. Clin. chim. Acta 113, 175-182 (1981).

23. Yamamoto, N., Nagasawa, Y., Sawai, M., Sudd, T., Tsubomura, H.: Potentiometric investigations of antigen-antibody and enzyme-enzyme inhibitor reactions using chemically modified metal electrodes. J. Immunol. Methods 22. 309-317 (1978).

24. Brown, M.J.: Monoclonals conjugated to electronic probe in automated assay system by Seattle company. Genetic Engineering News 4, 6 (1984).

25. Fraticelli, Y.M., Meyerhoff, M.E.: On-line gas dialyzer for automated enzymatic analysis with potentiometric ammonia detection. Anal. Chem. 55, 359-364 (1983).

26. Monroe, D.: Immobilization of biocatalyst on nitrocellulose in bioselective electrodes. BioTechniques 2, 302-304 (1984).

27. Ruoslahti, E.: Antigen-antibody interaction, antibody, affinity and dissociation of immune complexes. In: "Immunoadsorbents in Protein Purification", Ed. Ruoslahti, E., Scandinavian Journal of Immunology 3, 3-7 (1976).

THERAPEUTIC DRUG MONITORING BY AUTOMATED FLUORESCENCE POLARIZATION IMMUNOASSAY

D. Haidukewych
Epilepsy Center of Michigan, 3800 Woodward Avenue, Detroit, Michigan 48201, U.S.A.

Introduction

The principles of fluorescence polarization were first described in 1926 (1). However, it took nearly three decades before the technique was applied to biological systems (2), and its application to the antigen-antibody reaction was first described in 1961 (3). The principles and practice of fluorescence polarization and its application to biological systems have been the subject of several reviews (2, 4, 5).

Fluorescence polarization immunoassay (FPIA) makes use of competitive-binding assay principles, measuring tracer binding directly, without the need for a separation procedure. The specificity of an immunoassay is thereby combined with the speed and convenience of a homogeneous method. The manual use of FPIA has been applied to the measurement of such substances as insulin (6), cortisol (7), gentamicin (8, 9), and phenytoin (10).

Despite the advantage of FPIA and its potential as an assay technique, the method has seen little clinical application, primarily because of the lack of simple, high-performance instrumentation. There are several different instrumental designs (6, 11, 13), some of which are available commercially. The recent introduction of the fully automated Fluorescence Polarization Analyzer-TDx (Diagnostics Division, Abbott Laboratories, North Chicago, Illinois) has been hailed as the most significant breakthrough in the use of FPIA for routine therapeutic drug

Immunoassay Technology, Vol. 2

monitoring. In our laboratory, we have a choice of using either gas-liquid chromatography, homogeneous enzyme immunoassay (EMIT) or FPIA by TDx. As far as the stability of instrument calibration, superiority of precision, accuracy, speed and ease of use are concerned, FPIA is our method of choice.

This review on automated fluorescence polarization immunoassay will be limited to the use of TDx in therapeutic drug monitoring. The discussion will include theory of measurement, fluorescence polarization immunoassay, instrument design and operation, and clinically important assays for antiasthmatic, antiarrhythmic, antibiotic, anticonvulsant, antineoplastic, and cardiac glycoside drugs.

Theory of Fluorescence Polarization Measurement

Upon excitation by a light source, fluorescein (or its derivatives) absorbs blue light and emits green fluorescence after an excited-state lifetime of about four nanoseconds. If a molecule of fluorescein is excited by a plane-polarized beam of light, it will emit a polarized beam of light. The degree of polarization will be inversely related to how much the molecule has rotated between the time of excitation and emission. The time it takes for a molecule to rotate through an angle of 68.5 degrees has been defined as the rotational relaxation time of a molecule.

The rotational relaxation time of a molecule in solution is primarily dependant on its volume, so the observed polarization of fluorescence of a molecule in solution gives a direct indication of its size. Fluorescein molecules in solution are randomly oriented. The rotation of these small molecules between the time of absorption of excitation light and emission of fluorescence is large. The resulting polarization of fluorescence is therefore low, nearly zero. The size of the drug-fluorescein complex used as a tracer is small, so that a free tracer in solution also gives low polarization. This observed relationship between molecular size and polarization of fluorescence can be applied to the antigen-antibody reaction. The fluorescence polarization of a small tracer increases as the antigen is bound by antibody and assumes the rotational relaxation time of a large antibody molecule. This application of fluorescence polarization to the antigen-antibody reaction is the basis of fluorescence polarization immunoassay.

Fluorescence Polarization Immunoassay

Concentrations of therapeutic drugs in patient's biofluid can be determined using fluorescence polarization immunoassay. The analyte to be measured is labelled with fluorescein as a tracer. The tracer, when excited by linearly polarized light, emits fluorescence with a degree of polarization inversely related to its rate of rotation. Unbound tracer becomes essentially randomly oriented in the time between its absorption of excitation light and the emission of fluorescence. Upon binding to its specific antibody, the rotation of the tracer assumes that of the antibody molecule which is much slower than that of the relatively small tracer molecule. The polarization of emitted light increases when the tracer is bound. Unlabelled drug present in a patient's sample competes with the tracer for limited antibody sites and the observed polarization of fluorescence of the tracer becomes a value somewhere between that of the free and bound tracer. If the patient's sample contains a high concentration of the drug, the observed polarization value is closer to that of the free tracer, i.e. the value is low. If the patient's sample contains a low concentration of the drug, the polarization value is closer to that of the bound tracer, i.e. the value is high.

By sequentially exciting the reaction mixture of an immunoassay with vertically and then horizontally polarized light and analyzing only the vertical component of the emitted light, the polarization of fluorescence in the polarization mixture can be determined very accurately. The amount of polarization is calculated from the following equation (1) :

$$P = \frac{I_{vv} - I_{hv}}{I_{vv} + I_{hv}}$$

where I_{vv}= signal when vertically polarized light excites the sample and the intensity of the vertical component is analyzed. I_{hv}= signal when horizontally polarized light excites the sample and the intensity of the vertical component is analyzed. In the TDx, the polarization equation is modified to include correction for background intensity.

The precise relationship between polarization and concentration of the unlabelled drug is established by measuring the polarization values of calibrators with known concentrations of the drug. A nonlinear least-squares curve fit is used to interpolate the concentrations of unknown patient's sample.

Instrument Design and Operation

The TDx is a totally automated, bench-top, fluorescence polarization analyzer (14), in which all processing is carried out in a light-tight, temperature-controlled environment. The major components are a dual-syringe pump, liquid and air heaters, a pipettor boom, a polarization fluorimeter (15), microprocessor-based electronics, an alphanumeric printer, and a control panel. A 20-position carousel carries sample cups and cuvettes; tracer, antibody, and pretreatment reagents are contained in a three-chamber pack. Each pack has a bar-coded label that identifies the assay to the instrument. Buffer common to all assays is always on-line to the pump and is preheated immediately before delivery. The specimen (<100 µl) is placed in the sample side of the dual-chamber sample cup. A sample cup and cuvettes are required for each standard, control, or patient's specimen. Reagent packs for each assay are stored at 4-8°C, but can be loaded into the instrument without equilibration to room temperature. After loading the carousel and reagent pack, the processing is started by closing the door and pressing the run button.

The set-up is automatic. The bar-code on the reagent pack is read and processing parameters are retrieved from non-volatile memories. The number of samples to be processed is determined by an optical sensor, and all reagent volumes are checked with a liquid-sensor on the tip of the probe. The cuvettes are warmed to 35°C by an air heater controlled through an infrared sensor mounted in the optics module.

On the first revolution, samples are diluted with buffer in the dilution well of the sample cup. Typically, 20 µl of sample is mixed with 700 µl of buffer. The carousel then moves to the next sample position. On the second revolution, 10 to 25 µl (as required) of diluted sample and 25 µl of pretreatment solution are added to the cuvette along with buffer to a total volume of 975 µl. A blank reading is then made when the cuvette reaches the fluorimeter. On the third revolution of the carousel, 25 µl of tracer, 25 µl of antibody, and a volume of diluted sample equal to that used in the second revolution, are added with buffer to bring the total volume to 1.95 ml. After a 3-minute incubation, the final polarization measurement is made on the fourth revolution. A blank-corrected polarization value is calculated and either a standard curve is computed or the concentrations of specimens are determined by reference to a stored curve. A weighted log-logit curve fit is used. A hard-copy report is made with a 40-column printer. A full batch of 20 samples requires less than 10 minutes between loading samples and printout of the final results. A single determination is completed in less than 5 minutes. With the control panel, one can adjust the assays to be run and reported, changing replications, dilutions, and standard concentrations. Limits on reagent performance are stored for each assay, to check the integrity of samples and reagents. Numerous self-diagnostics are also provided for trouble-shooting problems and verifying system performance.

All functions of the instrument are under microprocessor

control. Incubation time, temperature, and pipetting are maintained to close tolerances. Under routine use in the clinical laboratory only one standard curve, which is stored in the computer, need be run per assay kit. Typically, in our laboratory, re-calibration (as judged by the high precision of control results) is not needed for at least several weeks.

Application of FPIA to Therapeutic Drug Monitoring

The review of automated FPIA as applicable to therapeutic drug monitoring will cover 23 presently available drug assays for TDx use: the antiasthmatic-theophylline; the antiarrhythmic drugs lidocaine, procainamide, N-acetylprocainamide (NAPA), quinidine and disopyramide; the antibiotics amikacin, gentamicin, kanamycin, netilmycin, streptomycin, tobramycin and vancomycin; the anticonvulsants carbamazepine, free carbamazepine, phenytoin, free phenytoin, phenobarbital, primidone and valproic acid; the antineoplastic methotrexate and the cardiac glycosides digoxin and digitoxin. Each drug will be considered separately by reviewing clinical indication, correlation between concentration in the biofluid and therapeutic-toxic effects, and correlation data on comparison of FPIA to such "reference" methods as enzyme immunoassay, radioimmunoassay (RIA), gas-liquid chromatography (GLC), high-pressure liquid chromatography (HPLC), substrate-labelled fluorescence immunoassay, and bioassays (Table 1, A to W).

The Antiasthmatic Theophylline

Theophylline (1,3-dimethylxanthine) is a naturally occurring compound with bronchodilator effect used in the treatment of asthma (16). Because of the narrow therapeutic index and the wide inter-individual variability in the rate of theophylline metabolism and clearance, nearly every patient receiving theophylline needs to have serum concentration determined. Correlations have been shown between serum levels and both therapeutic effect and toxicity. The data suggest that bronchodilator effects and bronchospasm inhibition (17) occur when serum theophylline concentrations are above 10 µg/ml. Adverse effects associated with concentrations above 20 µg/ml include nausea, headache and diarrhoea, and at higher levels, vomiting, gastrointestinal bleeding, seizures and cardiac arrhythmias (18). FPIA for theophylline has been described (19, 20) and comparison with RIA, enzyme immunoassay and HPLC is shown in Table 1A. Renal failure patients' specimens may show FPIA higher than HPLC values and, therefore,results must be confirmed by an alternative methodology.

Antiarrhythmic Drugs

Lidocaine

Lidocaine, administered intravenously, is a drug effective in the treatment of ventricular arrhythmias. It is also widely used as a local anaesthetic where significant systemic absorption can occur. Lidocaine therapeutic effectiveness and toxic side-effects show a consistent relationship to the blood concentration rather than to administered dosage. Because of the short half-life and the unpredictability of blood concentrations in patients suffering from cardiac or liver disease, lidocaine should be measured in patients to achieve effective therapy. Peak serum levels of lidocaine in the range of 1.5 to 5 µg/ml are suggested for optimal therapeutic effectiveness (21-23). Elevated concentrations of lidocaine (>10 µg/ml) may produce dizziness, tinnitus, respiratory arrest, and grand mal seizures. Patients in cardiac failure under long-term infusions may produce 50% greater plasma concentrations often resulting in central nervous system depression (24). Patients with chronic liver disease can have reduced lidocaine clearance and may develop toxicity with normal dosage regimens (25). Prophylactic administration of lidocaine is reported to produce instances of toxicity resulting in symptoms such as dizziness and nausea, and even coma or seizures (26, 27). Comparison of FPIA for lidocaine with Syva EMIT and HPLC is shown in Table 1B.

Procainamide

Procainamide is an effective and widely used antiarrhythmic drug. It has a narrow therapeutic index and can produce serious toxic side-effects (28-31). The relationship between the dosage and the resulting plasma concentration of procainamide has been found to be quite variable among patients. This variability is due to several factors: 1) Differences in drug metabolism among individuals. 2) Changes in the condition of the individual patient during therapy that alter drug metabolism. 3) Pathologic conditions such as renal impairment or cardiac failure, which have a profound effect on drug disposition. Therefore, monitoring the procainamide level in patient serum is recommended for safe and effective therapy. The commonly accepted therapeutic range for procainamide is 4-10 µg/ml. It is thought to be toxic at levels greater than 16 µg/ml. The presence of the active metabolite, N-acetylprocainamide (NAPA), creates problems in the interpretation of therapeutic drug monitoring information. It has been suggested that a total serum concentration of 5 µg/ml for procainamide plus NAPA is a reasonable lower limit for the therapeutic range and a total serum concentration of 20-30 µg/ml a reasonable upper limit (32). Further studies are required to more accurately determine the relationship of the combined procainamide and NAPA concentrations to therapeutic effect. FPIA for procainamide in comparison with enzyme immunoassay and HPLC is shown in Table 1C.

N-Acetylprocainamide (NAPA)

As indicated above, N-acetylprocainamide (NAPA) is the major active metabolite of procainamide. NAPA is considered comparable to the parent compound in activity and is present in equal or greater concentrations (31, 33). Levels of NAPA will vary widely. This variability is due to a half-life two to three times greater than that of procainamide, patients' renal status, and individual metabolism (34). Procainamide is metabolised to NAPA by N-acetyltransferase, an enzyme that is genetically determined and has biomodal distribution (32). The population is divided into "fast and slow acetylators". A fast acetylator with impaired renal function can have NAPA levels several times those of procainamide (35). The suggested therapeutic range for procainamide plus NAPA is 5-30 µg/ml, although presently there is no satisfactory way for interpreting the combined serum levels in relation to the therapeutic effect. FPIA for NAPA in comparison with enzyme immunoassay and HPLC is shown in Table 1D.

Quinidine

Quinidine is a naturally occurring alkaloid which is widely used to treat cardiac arrhythmias. It is given orally or parenterally as a sulphate, gluconate, or polygluconate form of the drug, and intravenously by administering quinidine gluconate by slow infusion (36). Quinidine is used to prevent both atrial and ventricular arrhythmias. Quinidine is bound to protein, principally albumin, to the extent of 70%-80%. Plasma protein binding of the drug is reduced in liver disease (37). Quinidine is metabolised in the liver. Principal metabolites, as found in serum, are 3-hydroxyquinidine, 2-oxoquinidine, and O-desmethylquinidine (38). 3-Hydroxyquinidine is reported to have an antiarrhythmic potency possibly equal to that of quinidine (39).

Many studies have shown a relationship between serum quinidine levels and its therapeutic effectiveness. Optimum therapeutic effects are usually observed when serum levels are between 2 to 5 µg/ml. At concentrations below 2 µg/ml, the patient generally experiences little relief from the symptoms of arrhythmia and above 5 µg/ml, toxic symptoms become evident. Mild side-effects include anorexia, nausea, vomiting, and diarrhoea. Toxic doses may produce what is termed "cinchonism". This includes gastrointestinal disturbances, tinnitus, hypotension, and cyanosis. This condition may lead to death unless it is recognised early and the quinidine treatment interrupted. FPIA measures total serum quinidine concentration, including the active metabolites. Comparison of FPIA with Syva EMIT, HPLC, and the double extraction fluorimetric method is shown in Table 1E.

Disopyramide

Disopyramide is an effective antiarrhythmic drug used especially for ventricular arrhythmias (40). It has a narrow therapeutic index and can produce serious toxic side-effects (41). The commonly accepted therapeutic range is 2-5 µg/ml. Disopyramide is thought to be toxic at levels greater than 7 µg/ml. Toxic side-effects include dry mouth, nose and throat, blurred vision, nausea, mental confusion and urinary hesitancy (41). Comparison of FPIA with Syva EMIT and HPLC is shown in Table 1F.

Antibiotics

Amikacin

Amikacin is effective in the treatment of serious gram-negative infections and is particularly useful in those involving strains resistant to all other aminoglycosides (42). Amikacin is probably the aminoglycoside of first choice when gentamicin resistance is strongly suspected (43). Patient samples which contain Kanamycin A, Kanamycin B, 3,4-dideoxykanamycin B or tobramycin will yield falsely elevated values for FPIA amikacin. However, these drugs are not usually administered together with amikacin. High concentration of penicillins or cephalosporins have been shown to inactivate aminoglycosides in vitro. Specimens from patients receiving additional antibiotics of these types should be assayed immediately or stored frozen. Peak serum levels of amikacin in the range of 20 to 25 µg/ml are suggested for optimum therapeutic effectiveness. Persistently elevated peak concentrations (30-35 µg/ml) have been shown to cause renal and central nervous system toxicity. Nephrotoxicity takes the form of damage to the proximal renal tubules and is associated with impaired renal function. Central nervous system toxicity is most often manifested as damage to the vestibular and auditory branches of the eight cranial nerve. Trough levels offer a more discrete indication of impending toxicity since they more closely correspond to tissue levels and are less affected by sampling errors (44). Comparison of FPIA with RIA and substrate-labelled fluorescence immunoassay is shown in Table 1G.

Gentamicin

Gentamicin is an aminoglycoside antibiotic which exhibits high potency and a broad spectrum bacterial action against both gram-negative and gram-positive organisms. It exhibits a narrow therapeutic index which makes its use hazardous, especially in patients with renal dysfunction. Therefore, accurate monitoring of the serum level in such patients is mandatory. In addition, the dose-serum level profile curve of gentamicin has been found to be surprisingly unpredictable, both in terms of peak-serum levels and elimination half-life from plasma (45). Patient samples which contain the drug sagamicin, sisomycin, and netilmicin will yield falsely elevated values for FPIA gentamicin. As is the case for amikacin, high concentration of penicillins or cephalosporins have been shown to inactivate gentamicin in vitro, therefore, specimens should either be assayed immediately or frozen.

Peak serum levels of gentamicin in the range of 5 to 10 µg/ml are suggested for optimal therapeutic effectiveness (45). Persistently elevated peak concentrations (>10 µg/ml) have been shown to cause renal and eight cranial nerve toxicity (46). Slowly rising trough levels have been shown to correspond to tissue accumulation of the drug, and trough levels >2 µg/ml have been associated with renal failure in some patients (44-46). FPIA for gentamicin has been described (14) and comparison with RIA, enzyme immunoassay, and substrate-labelled fluorescence immunoassay is shown in Table 1H.

Kanamycin

Kanamycin is clinically very effective in the neonatal period (47). It has a broad range of activity against gram-negative and gram-positive microorganisms. Similar to other aminoglycosides, kanamycin is ototoxic and may cause irreversible damage to the hair cells of the cochlea (47, 48). Excessive administration of kanamycin can also be associated with renal dysfunction (49). Trough concentrations >10 µg/ml and peak concentrations >35 µg/ml are considered toxic (49). Comparison of FPIA with RIA , radioenzymatic immunoassay, and bioassay is shown in Table 1I.

Netilmicin

Netilmicin is a new semi-synthetic aminoglycoside antibiotic which is a 1-N-ethyl derivative of sisomicin. Netilmicin has shown antimicrobial efficacy comparable to gentamicin, sisomicin, tobramycin, and amikacin and has also shown activity against many aminoglycoside-resistant bacteria (50-53). Netilmicin, similar to other aminoglycosides, is a potential ototoxic and nephrotoxic drug. Studies have shown that if strong consideration is given to renal function and to serum concentration of netilmicin, the ototoxic risk of netilmicin is low, probably lower than for gentamicin (54). Despite consistent dosage regimen, and normal renal function, a broad range of both peak and trough serum netilmicin levels has been found, as is also the case with gentamicin. The limited data at present suggests therapeutic efficacy of netilmicin at a trough level <2 µg/ml and at peak levels of 6 to 10 µg/ml (54-56). Comparison of FPIA with RIA, EMIT, and bioassay is shown in Table 1J.

Streptomycin

Streptomycin was the first aminoglycoside discovered, but its use has been limited by resistance of many strains of enteric bacteria and by discovery of newer agents with greater activity. Today it is used primarily in combination with other antimicrobial agents to treat serious enterococcal infections and tuberculosis (57). Although steptomycin is less nephrotoxic than other aminoglycosides, patients should have their serum streptomycin concentration monitored in order to assure that desired levels are attained and to avoid drug accumulation which might result in ototoxicity and nephrotoxicity (58). Desirable trough streptomycin levels should be 5 to 20 µg/ml. Toxicity has occurred at streptomycin concentrations of >40 to 50 µg/ml (57, 58). Comparison data of FPIA with bioassays is shown in Table 1K.

Tobramycin

Tobramycin is an aminoglycoside antibiotic whose in vitro activity is similar to that of gentamicin against gram-negative bacteria, including staphylococci. However, its activity against pseudomonas is higher than gentamicin (59). Tobramycin exhibits a narrow therapeutic index which makes its use hazardous, especially in patients with impaired renal function. Peak serum levels of tobramycin in the range of 5 to 10 µg/ml are suggested for optimal therapeutic effectiveness (60). Slowly rising trough levels have been shown to correspond to tissue accumulation of the drug, and levels >2 µg/ml have been associated with renal failure in some patients (44, 61). Comparison of FPIA with RIAs, enzyme immunoassay, and substrate-labelled fluorescence immunoassay is shown in Table 1L.

Vancomycin

Vancomycin is a glycopeptide antibiotic which is bactericidal against many gram-positive and some gram-negative cocci. It is useful in the therapy of severe staphylococcal (including methicillin-resistant staphylococci) infections in patients who cannot receive or who have failed to respond to the penicillins and cephalosporins. Vancomycin has been used successfully alone in the treatment of staphylococcal endocarditis (61, 62). Concurrent and sequential use of other neurotoxic or nephrotoxic antibiotics, requires careful monitoring of vancomycin (61-63). Peak serum levels of vancomycin in the range of 30 to 40 µg/ml and trough levels of 5 to 10 µg/ml are suggested for optimal therapeutic effect (63). In the presence of impaired renal function, unnecessarily high blood levels of vancomycin (90 µg/ml) may damage the eight cranial nerve and cause deafness (61, 63). FPIA for vancomycin has been described (64) and comparison with HPLC, RIA, and bioassay is shown in Table 1M.

Anticonvulsants

Carbamazepine

Carbamazepine is indicated for treatment of complex partial seizures, generalised tonic-clonic seizures, and elementary partial seizures (65). Carbamazepine is an iminostilbene derivative and is therefore structurally related to the tricyclic antidepressants. Toxicity associated with carbamazepine therapy is generally relatively minor. Plasma concentrations between 4 and 10 µg/ml have been associated with optimal seizure control in adults (66). Comparison of PFIA with enzyme immunoassay and HPLC methods is shown in Table 1N.

Free Carbamazepine

Carbamazepine in plasma is about 75% bound to albumin (67). Carbamazepine is metabolised extensively by the hepatic mixed-function oxidase system, yielding primarily the 10,11-epoxide which is quite stable, pharmacologically active, and found in the plasma and tissues. The 10,11-epoxide is further metabolised to the 10,11-dihydroxy derivative and eliminated in the urine as such or as its glucuronic acid conjugate (67). It is assumed that plasma carbamazepine levels are subject to considerable fluctuations due to the variable protein binding of the drug and its metabolite. Since only the unbound (protein-free) drug fraction is pharmacologically active, the use of total (bound plus free) plasma drug concentrations to adjust patient's dosage may be misleading. Sample preparation for FPIA of free carbamazepine requires separation of free carbamazepine from the protein-bound fraction prior to analysis. The classical equilibrium dialysis procedure is not applicable for routine use but ultrafiltration devices ready and convenient for use are commercially available.

The relationship between therapeutic effect and free levels of carbamazepine has not been established. Free carbamazepine levels have been reported to vary from 0.88 to 3.80 μg/ml (68). FPIA for free carbamazepine has been described, but the extensively variable protein-binding for carbamazepine reported in the literature could not be confirmed for "normal" epileptic outpatients (69). Comparison of FPIA with HPLC and Syva EMIT methods is shown in Table 10.

Phenytoin

Phenytoin is one of the most widely prescribed anticonvulsants and is occasionally used as a myocardial antiarrhythmic. In the treatment of epilepsy it is indicated for grand mal and psychomotor seizures (70). The main pathway (90%) for disposition of phenytoin is by excretion of the glucuronide of para-hydroxyphenylphenylhydantoin (HPPH) in the urine (71). It is hydroxylated in the liver and eliminated. The metabolic conversion to HPPH is a saturable process and in many cases small increments in dosage can cause a large increase in phenytoin plasma level (72). Because of the narrow therapeutic index and the wide inter-individual variability in the rate of phenytoin metabolism and clearance, the determination of serum levels of phenytoin for patients receiving therapy is essential.

Phenytoin toxicity affects primarily the central nervous system. Toxic levels can lead to nystagmus, vertigo, ataxia, psychoses, and even convulsions (71). Chronic treatment can lead to hyperplasia of gums (73), anaemia (74), and osteomalacia (75). The frequency and severity of dose-dependent toxic effects increases as the serum level rises above 20 µg/ml. Most patients will receive maximum seizure control at 10 to 20 µg/ml. FPIA for phenytoin has been described (76), and comparison with enzyme immunoassay, substrate-labelled fluorescence immunoassay and HPLC methods is shown in Table 1P.

Free Phenytoin

Due to the fact that phenytoin is highly bound to plasma protein (89%-95%) (77-79), any alteration in protein binding of phenytoin due to uraemia (80), hypoalbuminaemia (72), ingestion of other drugs, or age (81) can result in a significantly different clinical response to a particular total phenytoin concentration. When a patient's clinical response does not agree with the total phenytoin concentration, or the protein binding of a patient is believed to be abnormal, the free phenytoin level may correlate more accurately than the total level with clinical effectiveness or toxicity of the drug (77). In such cases, determination of the free phenytoin can be of value. As is the case for free carbamazepine, the free phenytoin in the patient's specimen must be separated from the bound phenytoin before performing the FPIA assay. Calibrators and controls do not require this separation step because all the drug is in the free state. The frequency and severity of dose-dependent toxic effects increases as free phenytoin rises to or above 3.0 µg/ml (77). Comparison of FPIA with HPLC and GLC methods is shown in Table 1Q.

Phenobarbital

Phenobarbital is useful in the treatment of epilepsy, especially for controlling focal motor, sensory, and grand mal seizures (82). Because of the narrow therapeutic index and the wide inter-individual variability in the rate of phenobarbital metabolism and clearance, the determination of blood levels of phenobarbital for patients receiving therapy is essential (83). Toxic levels of phenobarbital can lead to nystagmus, vertigo, and ataxia. A small number of patients develop hypersensitivity to the drug (84). Most patients will receive maximum seizure control when serum levels of phenobarbital are in the range of 15-40 µg/ml (85).

FPIA for phenobarbital cross-reacts 100% with parahydroxyphenobarbital. Therefore, for uremic patients alternative methodology should be used to confirm the results. FPIA for phenobarbital has been described (76) and comparison with enzyme immunoassay, substrate-labelled fluorescence immunoassay and HPLC methods is shown in Table 1R.

Primidone

Primidone is an anticonvulsant drug used in the treatment of complex partial seizures and their secondary generalisation (86). Primidone is metabolised in the liver to phenobarbital and phenylethyl-malonamide (86). In comparison to primidone, phenobarbital has a much longer half-life and therefore phenobarbital may accumulate to toxic levels in patients on primidone therapy (87, 88). Quantitation of phenobarbital is essential for patients on primidone therapy (89). Because of the narrow therapeutic range of primidone and the inter-patient variability in metabolism, measurement of primidone (and phenobarbital) is necessary for optimum therapy. The frequency of toxic symptoms (confusion, dizziness, nystagmus and ataxia) increases as primidone plasma levels approach 13-15 µg/ml (90). The therapeutic serum levels of primidone have been shown to range between 5 and 12 µg/ml (91). Comparison of FPIA with enzyme immunoassay, GLC, HPLC, and substrate-labelled fluorescence polarization is shown in Table 1S.

Valproic acid

Valproic acid is indicated for use both as the sole or adjunctive therapy in the treatment of simple (petit mal) and complex absence seizures. It may also be used adjunctively in patients with multiple seizure types including absence seizures (92). There is no precise relationship between valproic acid serum levels and controls of seizures (93), although most patients require >50 µg/ml for effective therapy (94). A therapeutic range of 50-100 µg/ml has been suggested (92). The main side-effects of valproic acid are gastrointestinal. Nausea, vomiting, and diarrhoea have been reported in up to 16% of adults and over 22% of children on initiation with valproic acid therapy (95). A rare adverse effect of valproic acid is hepatic dysfunction (94). Comparison of FPIA with enzyme immunoassay, HPLC and GLC is shown in Table 1T.

The Antineoplastic Methotrexate

Methotrexate is an antineoplastic drug used solely or in combination with other antineoplastic drugs for the treatment of leukaemia and other diseases (96). Severe psoriasis, sarcoidosis and granulomatosis have been treated with methotrexate in relatively low doses (97, 98). High dose methotrexate (>20 mg/kg weight) with citrovorum-factor rescue has been used with favourable results in the treatment of osteogenic sarcoma, leukaemia, non-Hodgkin's lymphoma, lung cancer, carcinoma of the head and neck, and breast cancer (99-104). The efficacy of methotrexate in the treatment of other tumours, such as prostatic cancer, is being investigated (105).

No precise relationship between serum levels of methotrexate and antineoplastic efficacy has been established, although levels below 2×10^{-8}M were seen as necessary for resumption of DNA synthesis (106). Following a 4-6 hour intravenous methotrexate infusion, a patient with a 24-hour serum concentration

>5 x 10^{-6}M to 10^{-5}M, a 48-hour level >5 x 10^{-7}M to 10^{-6}M, and a 72-hour level >10^{-7}M is at increased risk of toxicity if conventional low-dose leucovorin rescue is given. In this case, high-dose leucovorin rescue would be indicated (107-111). Toxicity typically presents in the form of myelosuppression, stomatitis, nausea, vomiting, convulsions, and liver and renal abnormalities (112). Comparison of FPIA with enzyme immunoassay at wide range of methotrexate concentrations is shown in Table 1U.

Cardiac Glycosides

Digoxin

Digoxin is a potent cardiac glycoside widely prescribed for the treatment of patients suffering from congestive heart failure, as well as some types of cardiac arrhythmias. Digoxin intoxication is a common and serious problem in the clinical setting. This is in part due to the fact that cardiac glycosides have a low therapeutic ratio (113). Coupled with a narrow therapeutic range is a marked patient variability in response to the same dosage of drug, resulting in often unpredictable drug serum levels (114). Intoxication symptoms are often indistinguishable from the original condition for which the drug was prescribed. It may not be immediately obvious whether the patient has been under- or over-dosed. Monitoring serum digoxin levels combined with other clinical data can provide the physician with useful information to aid in adjusting patient dosage, achieving optimal therapeutic effect while avoiding toxic dosage levels (115-116). Samples for digoxin assay should be drawn at least six hours after the last oral dose has been administered, by which time a steady state between serum digoxin concentration and myocardial digoxin concentration has been reached. Samples from patients receiving digitoxin or crude digitalis therapy will show falsely-elevated FPIA values for digoxin. A pretreatment step must be performed on each digoxin sample (calibrators, controls, and patient specimens) before testing. Serum normally contains fluorescent compounds bound to protein which would interfere with FPIA digoxin assay. The treatment consists of the addition of trichloroacetic acid to the serum sample to precipitate protein, followed by centrifugation to obtain clear supernatant. FPIA for digoxin is then performed on the sample supernatant.

Numerous studies have shown a relationship between serum digoxin levels and its concentration in myocardial and other tissues. Optimum therapeutic effects are usually observed when serum levels are in the range from 0.8 ng/ml to 2.0 ng/ml. Below 0.8 ng/ml the patient generally received little relief from symptoms, and above 2.0 ng/ml the patient may begin to experience discomforting intoxication symptoms (117). These symptoms may include gastrointestinal disturbances such as nausea, vomiting, and diarrhoea; nervous system disturbances manifested by blurred vision, headache, and general weakness; and cardiac arrhythmias and slowing of the pulse (116). Mean concentrations between 2.0-2.7 ng/ml (range up to 4.3 ng/ml), which may be associated with toxicity in adults, exhibit no signs of cardiac rhythm disturbances in young children (118). After two years of age, serum values for children more closely approximate those of adults. It is important to note that distinction between adequate digitalization and toxicity in patients cannot be made on the basis of digoxin concentration alone. Most studies show significant overlap between the toxic and non-toxic groups. Additional factors to consider when evaluating the correct therapeutic dosage for each patient are age, thyroid condition, acid-base balance, hypoxia, hypokalemia, renal function, and other clinical factors (119). Comparison of FPIA with RIA methods is shown in Table 1V.

Digitoxin

Digitoxin is a potent cardiac glycoside prescribed for the treatment of patients suffering from congestive heart failure as well as some types of cardiac arrhythmias. Of course, the factors discussed for digoxin also apply to digitoxin. Studies have shown a relationship between serum levels of digitoxin and its therapeutic effectiveness. Optimum therapeutic effects are usually observed when serum levels are in the range from 10.0 ng/ml to 30.0 ng/ml. At concentrations <10.0 ng/ml the patient generally receives little relief from symptoms, and at >30.0 ng/ ml the patients may begin to experience discomforting intoxication symptoms (120, 121). The accuracy of FPIA by correlation with RIA methods is shown in Table 1W.

Conclusions

From the foregoing review it is clear that the availability of automated fluorescence polarization immunoassay methodologies for measurement of clinically important drugs is an established fact.

In the past fifteen years, interest in quantification of substances in biofluids has been enormous. Laborious wet-chemistry procedures used by highly trained staff in research laboratories gave way to simple automated procedures which can now be used by personnel with minimum technical training. The choice of any one of the many methods available for therapeutic drug monitoring depends on expertise of laboratory personnel, existing equipment, and workload. Whatever the choice, there is no doubt that automated fluorescence polarization immunoassay will retain a prominent place in the routine clinical laboratory.

Table 1. Fluorescence Polarization Immunoassay (FPIA) Accuracy by correlation of clinical specimens to "reference methods"

FPIA DRUG vs.	N	Intercept	Slope	r
A. Theophylline				
RIA	38	1.476	1.022	0.935
EIA	97	0.016	0.939	0.977
HPLC	97	-0.091	0.979	0.985
B. Lidocaine				
SYVA EMIT	95	0.080	0.945	0.980
HPLC	55	0.096	0.975	0.944
C. Procainamide				
EIA	262	-0.183	1.009	0.990
HPLC	262	0.029	0.988	0.977
D. N-Acetylprocainamide				
EIA	96	-0.598	1.061	0.986
HPLC	118	-0.216	1.083	0.985
E. Quinidine				
SYVA EMIT	90	0.047	0.93	0.983
HPLC	107	0.224	0.92	0.983
DEFM	80	0.102	1.06	0.976
F. Disopyramide				
EMIT	102	0.248	0.978	0.984
HPLC	102	0.294	0.989	0.966
G. Amikacin				
RIA A	143	0.38	0.914	0.965
RIA B	162	-0.264	0.887	0.971
SLFIA	92	0.003	0.990	0.977
H Gentamicin				
RIA A	46	0.379	0.686	0.970
RIA B	38	-0.536	1.092	0.979
RIA C	153	-0.756	1.002	0.958
EIA	78	0.198	0.984	0.944
SLFIA	58	0.330	0.911	0.942
I. Kanamycin				
RIA	15	-0.263	0.895	0.998
REIA	145	-0.628	0.843	0.982
BIA	18	-1.579	1.144	0.990
J. Netilmicin				
RIA	98	-0.584	1.016	0.972
SYVA EMIT	89	-0.693	1.089	0.979
BIA	51	-0.166	0.892	0.924
K. Streptomycin				
BIA A	39	2.41	1.16	0.885
BIA B	40	0.295	1.00	0.995
L. Tobramycin				
RIA A	35	0.186	0.904	0.985
RIA B	70	-0.219	0.628	0.944
RIA C	142	-0.156	0.773	0.922
RIA D	84	0.154	0.741	0.982
EIA	170	-0.248	0.934	0.951
SLFIA	78	0.170	0.840	0.956

M. **Vancomycin**				
HPLC	98	3.043	1.090	0.980
RIA	98	1.666	1.036	0.957
BIA	98	2.782	0.871	0.907
N. **Carbamazepine**				
EIA	146	0.186	1.031	0.984
HPLC	39	-0.069	1.026	0.980
O. **Free Carbamazepine**				
HPLC	52	0.0775	1.01	0.968
SYVA EMIT	52	-0.0107	0.908	0.970
P. **Phenytoin**				
EIA	170	0.351	0.968	0.980
SLFIA	197	1.057	0.980	0.945
HPLC	26	0.846	1.009	0.992
Q. **Free Phenytoin**				
HPLC	26	0.059	1.056	0.973
GLC	51	0.051	1.038	0.977
R. **Phenobarbital**				
EIA	165	1.234	0.956	0.992
SLFIA	179	0.561	0.919	0.985
HPLC	37	2.274	0.986	0.970
S. **Primidone**				
EIA	103	0.190	0.970	0.990
GLC	106	0.360	1.040	0.970
HPLC	71	0.110	0.950	0.970
SLFIA	94	0.090	0.969	0.990
T. **Valproic Acid**				
EIA	115	0.334	0.982	0.962
HPLC	41	1.14	0.972	0.962
GLC	49	2.06	1.09	0.969
U. **Methotrexate**				
EIA (0 - 1uM/L)	56	0.94	-0.01	0.984
EIA (1 - 10uM/L)	40	0.88	0.26	0.995
EIA (10 - 100uM/L)	32	0.94	2.26	0.993
EIA (>100 uM/L)	20	0.95	2.13	0.994
EIA (0 - 1700 uM/L)	148	0.98	2.11	0.997
V. **Digoxin**				
RIA A	109	0.083	0.902	0.950
RIA B	118	0.148	0.910	0.959
RIA C	117	0.124	0.875	0.984
RIA D	114	-0.021	0.863	0.976
W. **Digitoxin**				
RIA A	178	0.729	1.060	0.967
RIA B	76	-0.476	1.014	0.927

continued.....

Legend to Table 1.

[a]Data summarized from TDx-System Assays Manual

N:number of patients medicated with the particular drug.

r:correlation coefficient.

RIA:radioimmunoassay.

EIA:enzyme immunoassay.

HPLC:high pressure liquid chromatography.

DEFM:double extraction fluorometric method.

SFLIA:substrate-labeled flourescence immunoassay.

REIA:radioenzymatic immunoassay.

BIA:bioassay.

GLC:gas-liquid chromatography.

References

1. Perrin, F.: Polarization de la lumière de fluorescence. Vie moyenne des molécules dans l'état excité. J. Phy. Radium 7, 390-401 (1926).
2. Weber, G.: Rotational Brownian motion and polarization of the fluorescence of solutions. Adv. Protein Chem. 8, 415-459 (1953).
3. Dandliker, W.B., Feigen, G.A.: Quantification of the antigen-antibody reaction by the polarization of fluorescence. Biochem. biophys. Res. Commun. 5, 299-304 (1961).
4. Dandliker, W.B., de Saussure, V.A.: Review article: Fluorescence polarization immunochemistry. Immunochemistry 7, 799-828 (1970).

5. Dandliker, W.B., Kelly, R.J., Dandliker, J., Sarquhar, J., Levin, J.: Fluorescence polarization immunoassay. Theory and experimental method. Immunochemistry 10, 219-227 (1973).

6. Spencer, R., Toledo, F., Yoss, N.: Design, construction and two applications for an automated flow-cell polarization fluorometer with digital read out: enzyme-inhibitor (anti-trypsin) assay and antigen-antibody (insulin-insulin anti-serum) assay. Clin. Chem. 19, 838-844 (1973).

7. Kobayashi, Y., Amitani, K., Watanabe, F., Miyai, K.: Fluorescence polarization immunoassay for cortisol. Clin. chim. Acta. 92, 241-247 (1979).

8. Watson, R.A.A., Landon, J., Shaw, E.J., Smith, D.S.: Polarization fluoroimmunoassay of gentamicin. Clin. chim. Acta 73, 51-55 (1976).

9. Jolley, M.E., Stroupe, S.D., Wang, C.J., Panas, H.N., Keegan, C.L., Schmidt, R.L., Schwenzer, K.S.: Fluorescence polarization immunoassay. I. Monitoring aminoglycoside antibiotics in serum or plasma. Clin. Chem. 27, 1190-1197 (1981).

10. McGregor, A.R., Crookall-Greening, J.O., Landon, J., Smith, D.S.: Polarization fluoroimmunoassay of phenytoin. Clin. chim. Acta 83, 11-166 (1978).

11. Jameson, D.M., Weber, G., Spencer, R.E., Mitchell, G.: Fluorescence polarization: Measurement with a photon counting photometer. Rev. Sci. Instrum. 49, 510-514 (1978).

12. Smith, J.C., Graham, N., Chance, B.: Fluorescence polarization detector for the computation of the degree of polarization. Rev. Sci. Instrum. 49, 1491-1492 (1978).

13. Kelly, R.J., Dandliker, W.B.: Digital photon-counting fluorescence polarometer. Anal. Chem. 48, 846-856 (1976).

14. Jolley, M.E., Stroupe, S.D., Schwenzer, K.S., Wang, C.J., Lu-Steffes, M., Hill, H.D., Popelka, S.R., Holen, J.T., Kelso, D.M.: Fluorescence polarization immunoassay. III. An automated system for therapeutic drug monitoring. Clin. Chem. 27, 1575-1579 (1981).

15. Popelka, S.R., Miller, D.M., Holen, J.T., Kelso, D.M.: Fluorescence polarization immunoassay. II. Analyser for rapid, precise measurement of fluorescence polarization with use of disposable cuvettes. Clin. Chem. 27, 1198-1201 (1981).

16. Hendeles, L., Weinberger, M.: Theophylline: "State of the art" review. Pharmacotherapy 3, 2-44 (1983).

17. Jenne, J., Wyze, E., Rook, F., MacDonald, F.M.: Pharmacokinetics of theophylline. Application to adjustment of the clinical dose of aminophylline. Clin. Pharmacol. Ther. 13, 349-360 (1972).

18. Jacobs, M.H., Senior, R.M., Kessler, G.: Clinical experience with theophylline. Relationships between dosage, serum concentration, and toxicity. JAMA 235, 1983-1986 (1976).

19. Hill, H.D., Jolley, M.E., Wang, C.H.J., Quille, C.J., Nystrom, D.D., Olive, D.L., Panas, H.N., Stroupe, S.D.: Fluorescence polarization immunoassay (FPIA) for theophylline; clinical correlation and reagent stability. Clin. Chem. 27, 1086 (1981).

20. George, H., Fisher, S., Horton, C., DiMarzio, J., Kurtz, S.: Simultaneous comparative evaluation of the optimate and TDx analysers using NCCLS guidelines. Clin. Chem. 30, 953 (1984).

21. Koch-Weser, J.: Correlation of serum concentrations and pharmacologic effects of antiarrhythmic drugs. Proceedings of the 5th International Congress of Pharmacology 3, 59-85 (1972).

22. Harrison, D.C., Alderman, E.L.: The pharmacology and clinical use of lidocaine as an antiarrhythmic drug. Modern Treatment 9, 139-175 (1972).

23. Baselt, R.C., Cravey, R.H.: A compendium of therapeutic and toxic concentrations of toxicologically significant drugs in human biofluids. J. Anal. Toxicol. 1, 81-103 (1977).

24. Thomson, P.D., Melmon, K.L., Richardson, J.A., Cohn, K., Steinbrunn, W., Cudiheerrowland, M.: Lidocaine pharmacokinetics in advanced heart failure, liver disease, and renal failure in humans. Ann. intern. Med. 78, 499-508 (1973).

25. Williams, R.L., Balschke, T.F., Meffin, P.J., Melman, K.L., Rowland, M.: Influence of viral hepatitis on the disposition of two compounds with high hepatic clearance: lidocaine and indocyanine green. Clin. Pharmacol. Ther. 20, 290-299 (1976).

26. Pitt, A., Lipp, H., Anderson, S.T.: Lignocaine given prophylactically to patients with acute myocardial infarction. Lancet 1, 612-616 (1971).

27. Mogenson, L.: Ventricular tachyarrhythmias and lignocaine prophylaxis in acute myocardial infarction. Acta med. Scand. (Suppl.) 513, 1-80 (1970).

28. Koch-Weser, J., Klein, S.W.: Procainamide dosage schedules, plasma concentrations and clinical effects. JAMA 215, 1454-1460 (1971).

29. Garfein, O.B.: Pharmacology of commonly used antiarrhythmic drugs and comments on the use of therapeutic drug monitoring. Ther. Drug Monitor. 4, 1-14 (1942).

30. Brown, J.E., Shand, D.G.: Therapeutic drug monitoring of antiarrhythmic agents. Clin. Pharmacokinet. 7, 125-128 (1982).

31. Karlsson, E.: Clinical pharmacokinetics of procainamide. Clin. Pharmacokinet. 3, 97-107 (1978).

32. Atkinson, A.J. Jr., Strong, J.M.: Effects of active drug metabolites on plasma level-response correlations. J. Pharmacokinet. Biopharm. 5, 95-109 (1977).

33. Atkinson, A.J. Jr., Stec, P.G., Lertora, J.J.L., Ruo, T.I., Thenot, J.P.: Impact of active metabolites on monitoring plasma concentrations of therapeutic drugs. In: "Frontiers in Therapeutic Drug Monitoring", Eds. Tognoni, G., Latini, R., Jusko, W.J., Raven Press, New York, pp. 17-25 (1980).

34. Brown, J.E., Shand, D.G.: Therapeutic drug monitoring of antiarrhythmic agents. Clin. Pharmacokinet. 7, 125-145 (1982).

35. Atkinson, A.J. Jr., Krumovsky, F.A., Huang, C.M., del Greco, F.: Hemodialysis for severe procainamide toxicity: Clinical and pharmacokinetic observations. Clin. Pharmacol. Ther. 20, 582-592 (1976).

36. Anderson, J.L., Harrison, D.C., Meffin, P.J., Winkle, R.A.: Antiarrhythmic drugs: Clinical pharmacology and therapeutic uses. Drugs 15, 271-309 (1978).

37. Reidenberg, M.M., Affrime, M.: Influence of disease on binding of drugs to plasma proteins. Ann. N.Y. Acad. Sci. 226, 115-126 (1973).

38. Drayer, D.E., Cook, C.E., Reidenberg, M.M.: Active quinidine metabolites. Clin. Res. 24, 623A (1976).

39. Drayer, D.E., Lowenthal, D.T., Restivo, K.M., Schwartz, A., Cook, C.E., Reidenberg, M.M.: Steady-state serum levels of quinidine and active metabolites in cardiac patients with varying degrees of renal function. Clin. Pharmacol. Ther. 24, 31-39 (1978).

40. Opie, L.H.: Drugs and the heart. IV. Antiarrhythmic agents. Lancet 1, 861-867 (1980).

41. Heel, R.C., Brogden, R.N., Speight, T.M., Avery, G.S.: Disopyramide: A review of its pharmacological properties and therapeutic use in treatment of cardiac arrhythmias. Drugs 15, 331-368 (1978).

42. Tally, F.P., Lovie, T.J., Weinstein, U.M., Bartlett, J.G., Gorbach, S.L.: Amikacin therapy for severe gram-negative sepsis, emphasis on infections with gentamicin-resistant organisms. Ann. intern. Med. 83, 484-488 (1975).

43. Brewer, N.S.: Antimicrobial agents. II. The aminoglycosides streptomycin, kanamycin, gentamicin, tobramycin, amikacin, neomycin. Mayo Clin. Proc. 52, 675-679 (1977).

44. Mangione, A., Schentag, J.J.: Therapeutic monitoring of aminoglycoside antibiotics: An approach. Ther. Drug Monitor. 2, 159-167 (1980).

45. Barza, M., Lauerman, M.: Why monitor serum levels of gentamicin ? Clin. Pharmacokinet. 3, 202-215 (1978).

46. Dayal, V.S., Smith, E.L., McCain, W.G.: Cochlear and vestibular gentamicin toxicity. A clinical study of systemic and topical usage. Arch. Otolaryngol. 100, 338-340 (1974).

47. Winkel, S., Bonding, P., Kildegard Larsen, P., Roosen, J.: Possible effects of kanamycin and incubation in newborn children with low birth weight. Acta paediat. Scand. 67, 709-715 (1978).

48. Davidson, S., Brish, M., Rein, N., Rubinstein, M., Rubenstein, E.: Ototoxicity in premature infants treated with kanamycin. Pediatrics 66, 479-480 (1980).

49. Maitra, S.K., Yoshikawa, T.T., Guze, L.B., Scholtz, M.C.: Determination of aminoglycoside antibiotics in biological fluids: A review. Clin. Chem. 25, 1361-1367 (1979).

50. Miller, G.H., Arcieri, G., Weinstein, M.J., Waitz, J.A.: Biological activity of netilmicin, a broad-spectrum semi-synthetic aminoglycoside antibiotic. Antimicrob. Agents Chemother. 10, 827 (1976).

51. Fu, K.P., Neu, H.C.: In vitro study of netilmicin compared with other aminoglycosides. Antimicrob. Agents Chemother. 10, 526 (1976).

52. Rahal, J.J. Jr., Simberkoff, M.S., Kagan, K., Maldover, N.H.: Bactericidal efficacy of Sch 20569 and amikacin against gentamicin-sensitive and resistant organisms. Antimicrob. Agents Chemother. 9, 595 (1976).

53. Watanakunakoru, C.: Comparative in vitro activity of Sch 20656 netilmicin, gentamicin, and tobramycin. Antimicrob. Agents Chemother. 10, 382 (1976).

54. Nordstrom, L., Christensson, P., Haeger, K., Juhlin, J., Tjornstrom, I., Wallmark, E.: Netilmicin: Clinical evaluation of efficacy and toxicity of a new aminoglycoside. J. Int. Med. Res. 7, 117 (1979).

55. Jahre, J.A., Fu, K.P., Neu, H.C.: Clinical evaluation of netilmicin therapy in serious infections. Am. J. Med. 66, 67-73 (1979).

56. Broughton, A., Strong, J.E., Pickering, L.K., Bodey, G.P.: Radioimmunoassay and radioenzymatic assay of netilmicin, a new aminoglycoside. Clin. Chem. 23, 1125 (1977).

57. Anhalt, J.P.: Antimicrobial assays. In: "Laboratory Procedures in Clinical Microbiology", Ed. Washington II, J.A., Springer-Verlag, New York, pp. 681-700 (1981).

58. Schwenzer, K.S., Anhalt, J.P.: Automated fluorescence polarization immunoassay for monitoring streptomycin. Antimicrob. Agents Chemother. 23, 683-687 (1983).

59. Brogden, R.N., Pinder, R.M., Sawyer, P.R., Speight, T.M., Avery, G.S.: Tobramycin: A review of its antibacterial and pharmacokinetic properties and therapeutic use. Drugs 12, 166-200 (1976).

60. Lew, M.: Interpretation of aminoglycoside serum levels. Hosp. Pharm. 14, 465-472 (1979).

61. Appel, G.B., Neu, H.C.: The nephrotoxicity of antimicrobial agents. New Engl. J. Med. 296, 722-728 (1977).

62. Cook, F.V., Farrar, W.W. Jr.: Vancomycin revisited. Ann. intern. Med. 88, 813-818 (1978).

63. Geraci, J.E.: Vancomycin. Mayo Clin. Proc. 52, 631-634 (1977).

64. Schwenzer, K.S., Chao-Heui, J.W., Anhalt, J.P.: Automated fluorescence polarization immunoassay for monitoring vancomycin. Ther. Drug Monit. 5, 341-345 (1983).

65. Cereghino, J.J.: Carbamazepine relation of plasma concentration to seizure control. In: "Antiepileptic Drugs", Eds. Woodbury, D.M., Penry, J.K., Pippenger, C.E., Raven Press, New York, 2nd Ed., pp. 507-519 (1982).

66. Troupin, A., Ojeman, L.M., Halpern, L., Dodrill, C., Wilkus, R., Friel, P., Feighl, P.: Carbamazepine: A double-blind comparison with phenytoin. Neurology 27, 511-519 (1977).

67. Lawless, L.M., De Monaco, H.J., Muido, L.R.: Protein binding of carbamazepine in epileptic patients. Neurology 32, 415-418 (1982).

68. MacKichan, J.J., Duffner, P.K., Cohen, M.E.: Salivary concentrations and plasma protein binding of carbamazepine and carbamazepine-10,11-epoxide in epileptic patients. Br. J. Clin. Pharmacol. 12, 31-37 (1981).

69. Haidukewych, D.: Comparison of fluorescence polarization immunoassay and enzyme immunoassay for free carbamazepine in serum or plasma ultrafiltrates from epileptic patients, Clin. Chem. 30, 1425-1426 (1984).

70. Buchthal, F., Lennox-Buchthal, M.A.: Diphenylhydantoin: Relation of anticonvulsant effect to concentration in serum. In: "Antiepileptic Drugs", Eds. Woodbury, D.M., Penry, J.K., Schmidt, P.R., Raven Press, New York, pp. 193-209 (1972).

71. Kutt, H., Winters, W., Kokenge, R., McDowell, F.: Diphenylhydantoin metabolism, blood levels and toxicity. Arch. Neurol. 11, 642-648 (1964).

72. Porter, R.J., Layzer, R.B.: Plasma albumin concentration and diphenylhydantoin binding in man. Arch. Neurol. 32, 298-303 (1975).

73. Hyman, G.A., Sommers, S.C.: The development of Hodgkin's disease and lymphoma during anticonvulsant therapy. Blood 28, 416-427 (1966).

74. De Vries, S.I.: Haematological aspects during treatment with anticonvulsant drugs. Epilepsia 6, 1-15 (1965).

75. Dent, C.E., Richens, A., Rowe, D.J.F., Stamp, T.C.B.: Osteomalacia with long term anticonvulsant therapy in epilepsy. Br. med. J. 4, 69-72 (1970).

76. Lu-Steffes, M., Jolley, M.E., Pittluck, G., Panas, H., Olive, D., Wang, C.H.J., Nystrom, D., Stroupe, S.: Fluorescence polarization immunoassays of phenytoin and phenobarbital. Clin. Chem. 27, 1093 (1981).

77. Booker, H.E., Darcey, B.: Serum concentration of free diphenylhydantoin and their relationship to clinical intoxication. Epilepsia 14, 177-184 (1973).

78. Anavekar, S.N., Saunders, R.H., Wardell, W.M.: Parotid and whole saliva in the prediction of serum total and free concentrations. Clin. Pharmacol. Ther. 24, 629-637 (1978).

79. Lund, L., Berlin, A., Lunde, P.K.M.: Plasma protein binding of diphenylhydantoin in patients with epilepsy. Clin. Pharmacol. Ther. 13, 196-200 (1972).

80. Kinniburgh, D.W., Boyd, N.D.: Phenytoin binding to partially purified albumin in renal disease. Clin. Pharmacol. Ther. 29, 203-210 (1981).

81. Rane, A., Lunde, P.K.M., Jalling, B., Yaffe, S.J., Sjoqvist, F.: Plasma protein binding of diphenylhydantoin in normal and hyperbilirubinaemic infants. J. Pediat. 78, 877-882 (1971).

82. Buchthal, F., Lennox-Buchthal, M.A.: Phenobarbital relation of serum concentration to control of seizures. In: "Antiepileptic Drugs", Eds. Woodbury, D.M., Penry, J.K., Schmidt, R.P., Raven Press, New York, pp. 335-343 (1972).

83. Waddell, W.J., Butler, T.C.: The distribution and excretion of phenobarbital. J. clin. Invest. 36, 1217-1226 (1957).

84. Svensmark, O., Buchthal, F.: Accumulation of phenobarbital in man. Epilepsia 4, 199-206 (1963).

85. Buchthal, F., Svensmark, O.: Serum concentrations of diphenylhydantoin (Phenytoin) and phenobarbital and their relation to therapeutic and toxic effects. Psychiat. Neurol. Neurochir. 74, 117-136 (1971).

86. Schottelius, D.D.: Primidone: Bio-transformation. In: "Anti-epileptic Drugs", Eds. Woodbury, D.M., Penry, J.K., Pippenger, C.E., Raven Press, New York, 2nd Ed., pp. 415-420 (1982).

87. Penry, J.K., Nemark, M.E.: The use of antiepileptic drugs. Ann. intern. Med. 90, 207-218 (1979).

88. Conomy, J.P.: Long term use of the major anticonvulsant drugs. Am. Fam. Physician 18, 107-116 (1978).

89. Haidukewych, D., Rodin, E.A.: Monitoring 2-ethyl-2-phenylmalonamide in serum by gas-liquid chromatography: Application to retrospective study in epilepsy patients dosed with primidone. Clin. Chem. 26, 1537-1539 (1980).

90. Cloyd, J.C., Miller, K.W., Seppick, I.E.: Primidone kinetics: Effects of concurrent drugs and duration of therapy. Clin. Pharmacol. Ther. 29, 402-407 (1981).

91. Heipertz, R., Guthoff, A., Bernhardt, W.: Primidone metabolism in renal insufficiency and acute intoxication. J. Neurol. 221, 101-104 (1979).

92. Simon, D., Penry, J.K.: Sodium-di-n-propyacetate (DPA) in the treatment of epilepsy. Epilepsia 16, 549-573 (1975).

93. Lewis, J.R.: Valproic acid (depakene): A new anticonvulsant agent. JAMA 240, 2190-2192 (1978).

94. Browne, T.R.: Valproic acid. New Engl. J. Med. 302, 661-666 (1980).

95. Adams, D.J., Luders, H., Pippenger, C.: Sodium valproate in the treatment of intractable seizure disorders: A clinical and electroencephalographic study. Neurology 28, 152-157 (1978).

96. Goldin, A., Venditti, J.M., Kline, I.: Eradication of leukemia cells (L 1210) by methotrexate and methotrexate plus citrovorum. Nature 212, 1548-1550 (1966).

97. Myfors, A.: Methotrexate therapy of psoriasis. Dan. med. J. 27, 74-96 (1980).

98. Bleyer, W.A.: The clinical pharmacology of methotrexate. New application of an old drug. Cancer 41, 36-51 (1978).

99. Jaffe, N., Frei, E., Watts, H., Traggis, D.: High-dose methotrexate in osteogenic sarcoma: A 5-year experience. Cancer Treat. Rep. 62, 259-264 (1978).

100. Freeman, A.I., Weinberg, V., Brecher, M.L., Jones, B., Glicksman, A.S., Sinks, L.F., Weil, M., Pleuss, H., Hananian, J., Burgert, O. Jr., Gilchrist, G.S., Necheles, T., Harris, M., Kung, F., Patterson, R.B., Maurer, H., Leventhal, B., Chevalier, L., Forman, E., Holland, J.F.: Comparison of intermediate-dose methotrexate with cranial irradiation for the post-induction treatment of acute lymphocytic leukemia in children. New Engl. J. Med. 308, 477-484 (1983).

101. Djerassi, I., Kim, J.S.: Methotrexate and citrovorum factor rescue in the management of childhood lymphosarcoma and reticulum cell sarcoma (non-Hodgkin's lymphoma). Cancer 38, 1043-1051 (1976).

102. Djerassi, I., Rominger, J., Kim, J.S., Turchi, J., Suvansri, U., Hughes, D.: Phase I study of methotrexate with citrovorum factor in patients with lung cancer. Cancer 30, 22-30 (1972).

103. Vogl, S.E., Kaplan, B.H.: Chemotherapy of advanced head and neck cancer with methotrexate, bleomycin, and cis-diammine-dichloroplatinum II on an effective outpatient schedule. Cancer 44, 26-31 (1979).

104. Vogler, W.R., Jacob, J.: High divided dose methotrexate vs. single dose. IV. Methotrexate in advanced carcinoma of head, neck, breast and colon. Proceedings of the American Association of Cancer Research 16, 266 (1975).

105. Herr, H.: Cyclophosphamide, methotrexate and 5-fluorouracil combination chemotherapy versus chloroethyl-cyclohexyl-nitrosourea in the treatment of metastatic prostatic cancer. J. Urol. 127, 462-465 (1982).

106. Young, R.C., Chabner, B.A.: An in vivo method for monitoring differential effects of chemotherapy on target tissue in man: Correlation with plasma pharmacokinetics. J. clin. Invest. 52, 92 (1973).

107. Tattersall, M.H.N., Parker, L.M.: Clinical pharmacology of high-dose methotrexate. Cancer Chemother. Rep. 6, 25-29 (1975).

108. Isacoff, W.H., Townsend, C.M., Eilber, F.R.: High-dose methotrexate therapy of solid tumors: Observations relating to toxicity. Md. Pediatr. Oncol. 2, 319-325 (1976).

109. Isacoff, W.H., Morrison, P.F., Aroesty, J.: Pharmacokinetics of high-dose methotrexate with citrovorum factor rescue. Cancer Treat. Rep. 61, 1665-1674 (1977).

110. Nirenberg, A., Mosende, C., Mehta, B.M.: High-dose methotrexate with CF rescue: Predictive value of serum methotrexate concentrations and corrective measures to avert toxicity. Cancer Treat. Rep. 61, 779-783 (1977).

111. Evans, W.E., Pratt, C.B., Taylor, R.H.: Pharmacokinetic monitoring of high-dose methotrexate: early recognition of high-risk patients. Cancer Chemother. Pharmacol. 3, 161-166 (1979).

112. Jaffe, N., Traggis, D.: Toxicity of high-dose methotrexate (NSC-749) and citrovorum factor (NSC 3540) in osteogenic sarcoma. Cancer Chemother. Rep. 6, 31-36 (1975).

113. Bresnahan, J.F., Vliestra, R.E.: Series of pharmacology in practice. 3. Digitalis glycosides. Mayo Clin. Proc. 54, 675-684 (1979).

114. Surawicz, B., Mortelmans, S.: Factors affecting individual tolerance to digitalis. In: "Digitalis", Eds. Fisch, C., Suawicz, B., Grune and Stratton, New York, pp. 127-147 (1969).

115. McNeely, M.D.D.: Making digitalis safer. Drug Ther. 5, 222-224 (1975).

116. Physicians' Desk Reference: Ornadell, N.J.: Medical Economics Inc., 34th Edition, pp. 787-792 (1982).

117. Taggart, A.J., McDevitt, A.G.: Digitalis: Its place in modern therapy. Drugs 20, 398-404 (1980).

118. Butler, V.P. Jr.: Assays of digitalis in the blood. Progr. Cardiovasc. Dis. XIV, 571-600 (1972).

119. Huffman, D.H., Crow, J.W., Pentikainen, P., Azarnoff, D.L.: Association between clinical cardiac status, laboratory parameters and digoxin usage. Am. Heart J. 91, 28-34 (1976).

120. Moe, G.K., Farah, A.E.: Digitalis and allied cardiac glycosides. In: "The Pharmacological Basis of Therapeutics", Eds. Goodman, L.S., Gilman, A., Macmillan Publishing Co., New York, 5th Ed., pp. 653-682 (1975).

121. Beller, G.A., Smith, T.W., Abelmann, W.H., Haber, E., Hood, W.B.: Digitalis intoxication. New Engl. J. Med. 284, 989-997 (1971).

SPECIFIC ANTIBODY SYNTHESIS IN VITRO : AN APPRAISAL OF THE MICROCULTURE ANTIBODY SYNTHESIS ENZYME-LINKED ASSAY (MASELA)

A. L. Moore and W. B. Ershler
Vermont Regional Cancer Center and Department of Medicine,
University of Vermont, Burlington, Vermont, U.S.A.

Introduction

The haemolytic plaque assay (1) has provided a method for *in vitro* assessment of immunoglobulin secretion at the single cell level, while modifications of the assay, which involve coupling of specific antibodies or antigens to indicator erythrocytes, have significantly extended its applicability. Plaque forming cells (PFC) have been demonstrated in both human peripheral blood (2) and in organ derived lymphocyte populations (3). It was demonstrated (4) that human peripheral blood mononuclear cells (PBM) from actively immunized volunteers produce a vigorous PFC response to the immunogen Salmonella lipopolysaccharide (LPS). Plaque forming assays that require the preparation of stable conjugate red cells can be cumbersome, and are difficult to apply to a number of soluble antigens.

Specific antibody produced in cultures of PBM from immunized individuals has been measured in culture supernatants using a phage neutralizing assay (5), radioimmunoassay (6), or enzyme linked immunosorbent assay (ELISA) in polyclonal mitogen (7) or specific antigen stimulated cultures (8-10). These assays generally require two steps: the culture, followed by an antibody assay of the culture supernatant. A modified ELISA has been used to measure specific antibody secretion by short term cultures (1-20 hr) of immune murine or guina pig cells in antigen coated wells (11-13). Recently, a solid-phase Enzyme-Linked Immunospot

Immunoassay Technology, Vol. 2

(ELISPOT) assay has been described to enumerate immune mouse spleen cells secreting specific antibody (14). This technique, like the plaque assay, does not allow for the study of *in vitro* cell interactions or manipulations, but yields information regarding single cells. For many purposes, however, the total amount of antibody released may be of more biological importance than the frequency of cells.

We have described an assay whereby human PBM cells from recently immunized subjects are cultured in sterile antigen coated wells; the specific antibody that is synthesized adheres to the solid phase antigen and is measured in an ELISA. We have used this microculture antibody synthesis enzyme-linked assay (MASELA) to study *in vitro* antibody response in subjects recently vaccinated with tetanus toxoid (TT) (15-17) and influenza (18). This technique allows for the *in vitro* assessment of immune competence, the dissection of cellular interactions, and the evaluation of agents purported to influence basic immune functions either as enhancers or as inhibitors.

MASELA: Methodologic Aspects

Heparinized peripheral blood from individuals, pre- and 7 days post-vaccination, was separated by density gradient centrifugation on Ficoll/Hypaque. The mononuclear fraction (PBM) from the interface was washed three times and resuspended in complete media as described (16-19) and then cultured in the MASELA as outlined in Fig. 1.

Triplicates of 25, 50, 100 or 200 x 10^3 cells in 250 µl of complete media are cultured. Control cultures are established in wells that are not coated with antigen, or are coated with irrelevant antigen. To quantify the *in vitro* antibody production we have adapted a method described to measure the change in plasma antibody in response to influenza vaccine (Fig. 2)(18, 20). The pre- and post-immunization optical densities observed at

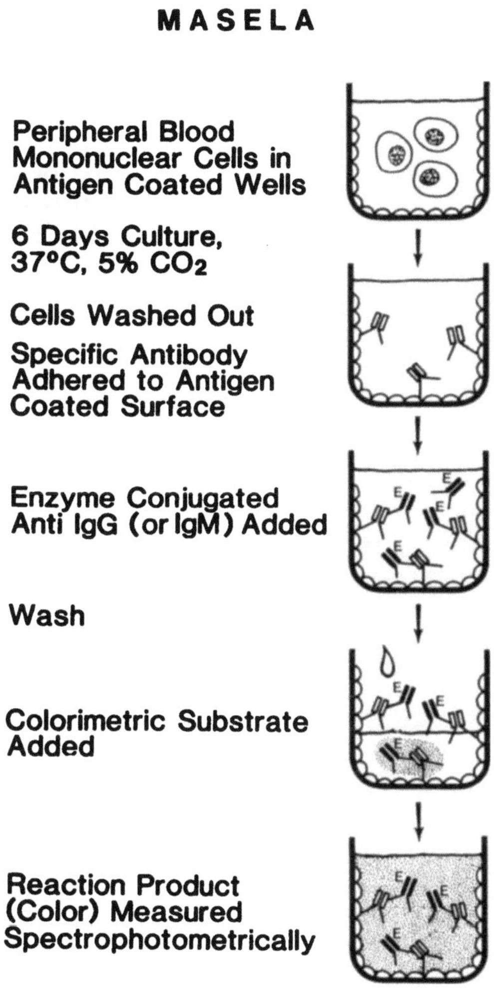

Fig. 1. Microculture antibody synthesis enzyme-linked assay (MASELA) for the determination of specific antibody synthesis *in vitro*. (Reference 25, with permission from the publisher).

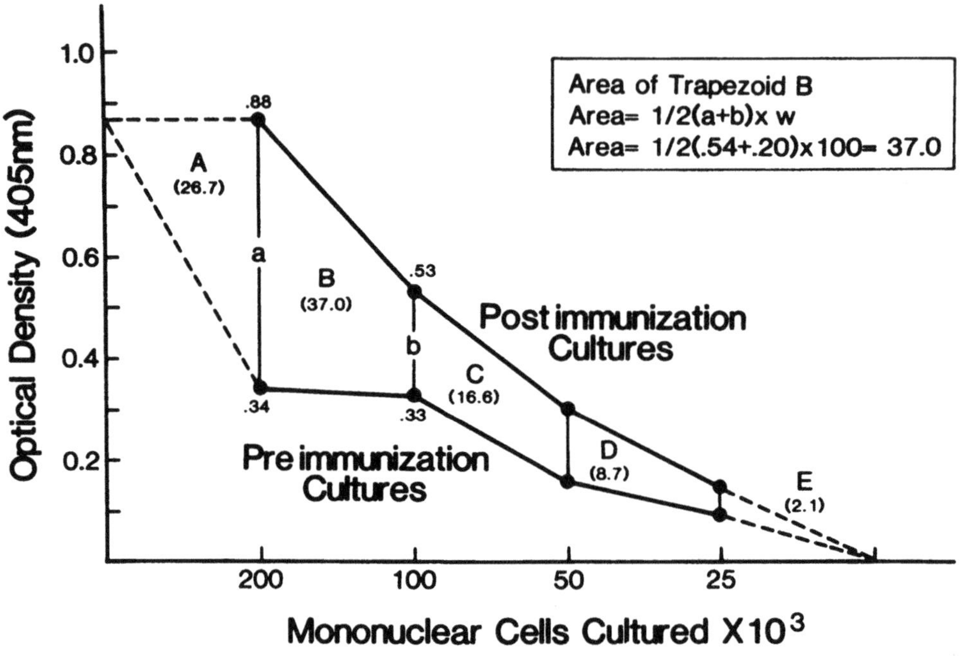

Fig. 2. Demonstration of antibody response in the microculture assay by determination of the area between pre- and post-immunization curves (modified from the method of Murphy et al., 20). Prior to and 1 week after immunization with the trivalent influenza vaccine, lymphocytes were cultured in the MASELA and the results for one individual are depicted. The areas were calculated, using the trapezoid rule by which the area is calculated according to the formula 1/2(a + b)w as indicated; w. (cell concentration) is a constant and is arbitrarily assigned a value of 100. The area between pre- and post-immunization cultures depicted in the graph is the sum of the individual areas A + B + C + D + E = 26.7 + 16.6 + 8.7 + 2.1 = 91.1. Areas B, C and D are measured areas, i.e. they are described between the starting concentration of 200 x 10^3 cells/culture and the most dilute concentration of 25 x 10^3 cells/culture. The areas of A and E are those generated by the hypothetical closure of the 2 curves. (Reference 18, with permission from the publisher).

each of four cell concentrations are plotted and the areas between the curves are estimated by the summation of the areas of the trapezoids and triangles between the curves. This area, therefore, represents the boost of in vitro antibody synthesis after vaccination.

To determine that the antibody measured in the MASELA is synthesized in vitro, and not merely passively released preformed antibody, immune PBM were either lysed by the freeze/thaw technique or incubated with puromycin (to inhibit protein synthesis) prior to culturing in the MASELA (Fig. 3). These manoeuvres markedly diminished the specific antibody present and indicate that viable PBM capable of protein synthesis are required (8, 16).

In this in vitro system, specific antibody production is greatest in cultures established 7 days post-booster immunization and returns to background levels when the cultures are established at 14 days (Fig. 4) (8). This corroborates findings that a PFC response developed in S. typhii LPS immunized individuals by day 7, peaked at day 10 and declined to baseline levels by day 14 (4).

Murine spleen cells primed in vitro with soluble antigen have been shown to produce detectable antibody in a modified ELISA (11, 12). Human PBM stimulated in vitro with live or inactivated influenza virus synthesized anti-influenza virus antibody (8, 9). Similarly, cultures of human PBM stimulated with the antigen Keyhole Limpet Haemocyanin have been shown to produce polyclonal immunoglobulin (21). In an effort to develop a system in which the specific in vitro antibody could be studied without the necessity of vaccinating individuals, we have stimulated PBM in vitro with influenza antigens and then cultured these sensitized PBM and measured specific antibody using the MASELA.

It has been shown that, under optimum conditions for in vitro Ag stimulation of antibody response culturing 2 million PBM for 12 days in flat bottom wells, antibody is produced by the progeny of only a few B cell precursors (8). Using similar

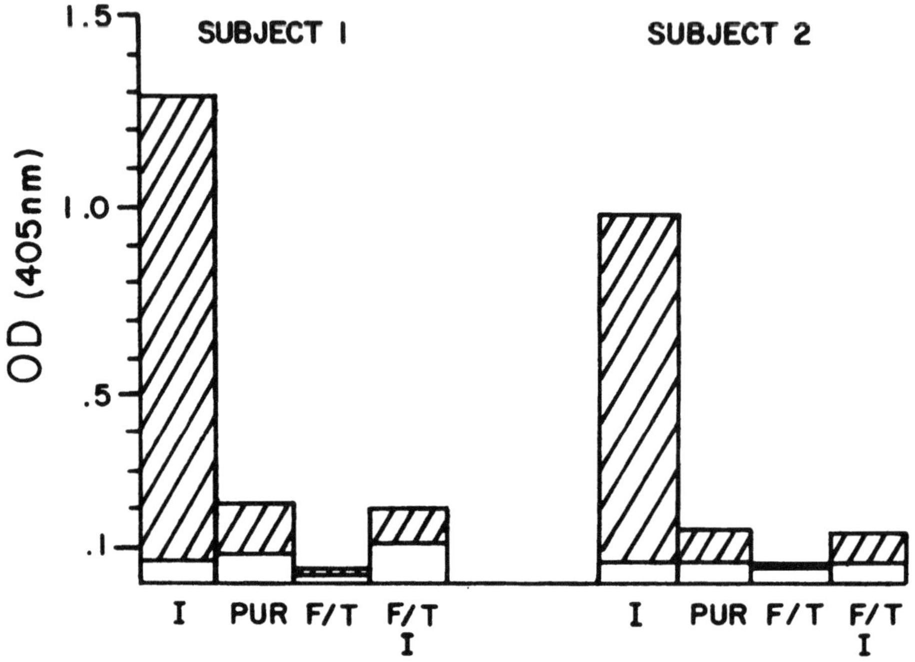

Fig. 3. The effects of cell disruption or inhibition of protein synthesis on production of antibody in the microculture ELISA. Peripheral blood mononuclear cells were either incubated (I) under standard MASELA conditions, frozen, and thawed, and then incubated (FT/I), or were frozen for the entire culture period and then thawed for antibody determination (FT). In certain wells, on the standard (not frozen) plate, protein synthesis was inhibited by puromycin (PUR). Anti-TT antibody produced in cultures established prior to booster immunization (open portion of bar) and that produced in cultures established 1 week after booster immunization (hatched plus open portion of bar) are depicted for 2 young volunteers. As can be seen, freeze/thaw and puromycin treatments abrogate the increment in antibody production seen in cultures established 1 week after booster immunization. (Reference 16, with permission from the publisher).

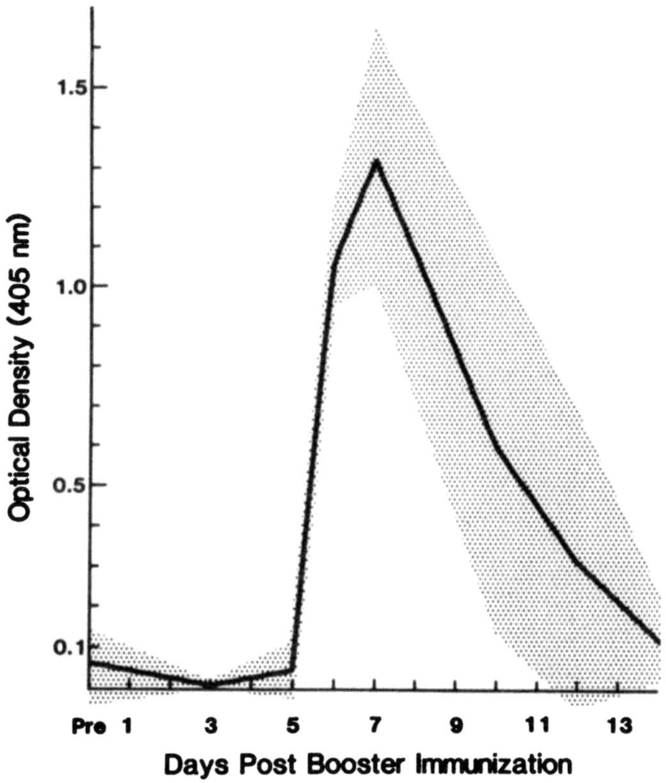

Fig. 4. *In vitro* anti-TT produced in lymphocyte cultures established after booster immunization. Lymphocytes taken from 11 different individuals were cultured (in quadruplicate) in antigen-coated wells prior to and at different days post-booster immunization. Antibody produced that adhered to the antigen was measured by an enzyme-linked immunosorbent assay. The solid line represents the mean optical density (shaded area: ± standard error) measured for several different individuals (n=11 for pre-immunization and day 14; n=4 for days 3, 5, 10 and 12; n=2 for day 6; and n=9 for day 7). As can be seen, the cultures established from blood samples obtained 7 days after booster immunization gave maximal production and by day 14, antibody production fell to pre-immunization levels. Control cultures in wells without initial TT coat or established from samples obtained from non-booster immunized individuals consistently resulted in optical densities (405 nm) of 0.15 or less. (Reference 16, with permission from the publisher).

conditions, but culturing PBM with influenza antigen for only 5 days, we were able to sensitize in vitro the PBM from 10 out of 20 individuals and these were shown to synthesize anti-influenza antibody in the MASELA. However, this in vitro immunization did not correlate with the in vivo antibody level measured in the plasma of these same 20 individuals post-vaccination (18). A longer incubation period with antigen (12 days) may have shown a better correlation.

Potential Applications of MASELA

a) Immune competence

The MASELA technique used to measure specific antibody response to immunization may prove to be a useful reflection of in vivo antibody response. We have found a positive correlation (r=0.81) (Fig. 5) between capabilities of in vitro antibody response and the in vivo plasma antibody levels after influenza vaccine.

Our interest in age acquired immune deficiency led us to investigate the in vitro and in vivo antibody response to influenza vaccine. The U.S. Public Health Service recommends annual influenza vaccine for all people over 65 years of age (22). Nevertheless, in some elderly people, deficient antibody responses have been demonstrated and influenza has clearly occurred in vaccinated subjects (23). In our study, we have shown that specific antibody response to the trivalent influenza vaccine was significantly less in a group of fully ambulatory, healthy, elderly volunteers when compared with young controls. This was true even when accounting for that component of the reduced response due to previous immunization and pre-immunization antibody levels. After a fairly extensive in vitro evaluation of immune competence, we found that the best predictor of antibody response was the in vitro specific antibody synthesis measured in the MASELA (19).

A similar approach could be used to study other clinically

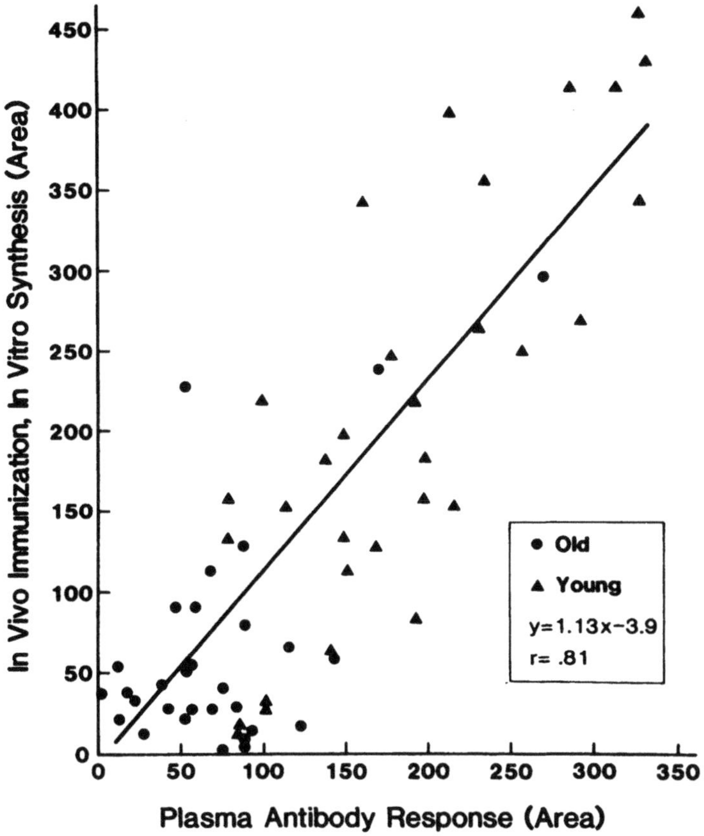

Fig. 5. Correlation between plasma antibody levels and _in vitro_ antibody synthesis after vaccination. Antibody response was determined by calculating the area between the pre- and post-immunization dilution curves and is reported in arbitrary area units (20). Greater production _in vitro_ was generally observed in individuals who responded with a higher plasma antibody level as well ($p<0.01$).

significant immune deficient states in which immunizations are recommended. Examples of these include patients with lymphoreticular malignancies, AIDS, post-splenectomy, trauma or burns.

b) Cellular interactions

One advantage of cell culture/antibody synthesis type assays compared with single cell release assays, such as PFC or ELISPOT assays, is that they allow for the dissection of cellular interactions. PBM can be fractionated into sub-sets and then cultured alone or in various combinations to determine the contribution of different cellular sub-sets. Antibody response to large antigens frequently requires a complex network of cellular interactions. The antibody response to such antigens as tetanus toxoid and influenza virus has been shown to be T-cell dependent (24).

Cell separation and remixture experiments have demonstrated that such interactions are in fact taking place in assays measuring _in vitro_ antibody synthesis. For example, helper factors from the supernatant of _in vitro_ primed T-lymphocytes specifically enhanced antibody synthesis from primed mouse spleen cells (12). Also, human PBM separated into B and T enriched subpopulations produced little _in vitro_ antibody to influenza virus. However, when cocultured the antibody response was reconstituted, indicating that the anti-influenza antibody production by B cells was T cell dependent (8). Similarly, production of anti-TT antibody in the MASELA requires T cells for optimal antibody response (Fig. 6) (16).

In these experiments, monocyte depleted human PBM did not decrease antibody response; however, T cell enriched cultures produced minimal anti-TT. B cell enriched cultures produced 50% less anti-TT, but when cocultured with increasing ratios of T cell enriched PBM produced increasing amounts of anti-TT.

c) Effect of biological response modifiers

The utilization of this assay with biological response modifiers, such as thymic hormones or lymphokines, particularly with purified cell populations or even cloned lymphocytes, may prove useful in understanding the complex network of cellular interactions. The importance of specific cellular components, or humoral factors (e.g. lymphokines) could be readily assessed.

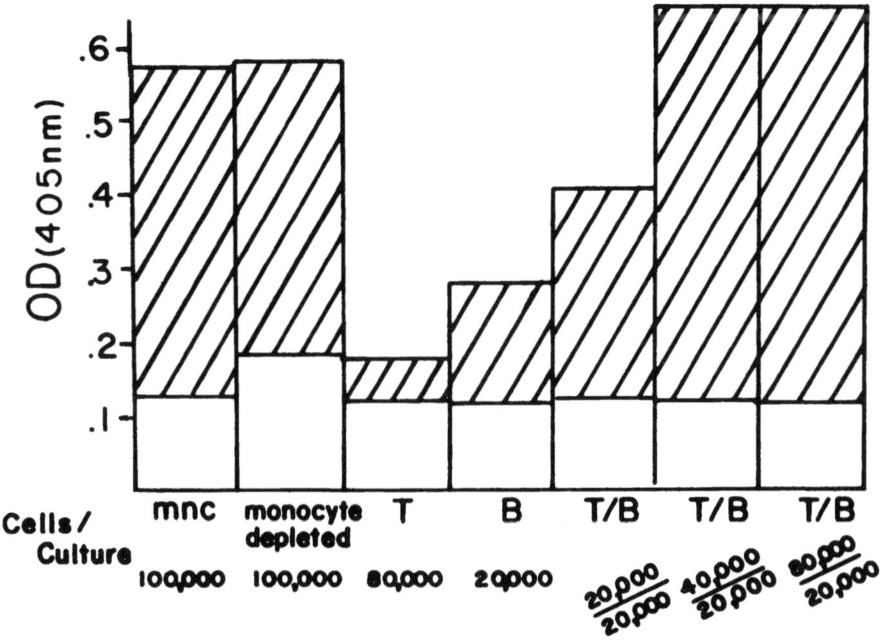

Fig. 6. Cellular interactions in the microculture ELISA. Peripheral blood mononuclear cells were fractionated into subpopulations by standard cell separation techniques. MD-MNC (monocyte depleted mononuclear cells); T (T cell enriched); and B (T cell and monocyte depleted). Cells were added to the culture in varying ratios and anti-TT IgG antibody production was measured in cultures established prior to (open portion of bar) and 1 week after (hatched plus open portion of bar) TT booster immunization. Production of anti-TT antibody by T cells alone was minimal, but the addition of the fraction augmented, in a dose-related way, antibody production from the 7-day (but not the pre-immunization) cultures of the B cell fractions. (Reference 16, with permission from the publisher).

From a clinical perspective, however, it would be necessary to show that <u>in vitro</u> antibody synthesis (i.e. in the MASELA) correlates well with <u>in vivo</u> antibody response, (i.e. antibody levels). If this is the case, as has already been demonstrated for influenza vaccine (Fig. 5), it would be reasonable to use this assay to test the augmenting capabilities of biological response modifiers and project the potential clinical efficacy of such intervention.

Accordingly, from our studies it is apparent that in vitro antibody production was less in those people in whom peak plasma levels were also less. It would also appear that it is in these same individuals that thymic hormones have their greatest effect (17, 19). This in vitro result was offered as a pre-clinical rationale for a trial of thymic hormone in elderly individuals who responded weakly to influenza vaccine, but whose in vitro antibody response was augmented significantly by coculture with the thymic hormone. Similar studies with other biological response modifiers would be of great interest.

d) Assessment of immunotoxicity of different agents

As mentioned above, the antibody response to certain antigens is complex, requiring the interaction of different cellular subsets. Experiments have supported the contentions that these cellular interactions are operant also in vitro, in the MASELA. It would, therefore, be possible to assess the effect of drugs or toxins on the antibody response in order to predict potential immunologic toxicity.

Conclusion

Immunological assays have been developed which measure the capacity for lymphocytes to synthesize and secrete immunoglobulins or specific antibody. One such assay, the MASELA, incorporates cell culture and ELISA technologies and provides information regarding specific antibody production after in vivo immunization. This technique can be applied to experimental animals and humans and may be useful in assessment of immune competence. Furthermore, the MASELA may be useful in analyzing the contribution of various lymphocyte sub-sets in the antibody response and possibly in determining the potential importance of immuno-modulators such as thymosin, lymphokines or immuno-active drugs.

References

1. Jerne, N.K., Nordin, A.D.: Plaque formation in agar by single antibody-producing cells. Science 140, 405 (1963).
2. Eby, W.C., Chong, C.A., Dray, S., Molinaro, G.A.: Enumerating immunoglobulin-secreting cells among peripheral human lymphocytes. A hemolytic plaque assay for a B cell function. J. Immunol. 115, 1700-1703 (1975).
3. Weetman, A.P., McGregor, A.M., Lazarus, J.H., Hall, R.: Thyroid antibodies are produced by thyroid-derived lymphocytes. Clin. Exp. Immunol. 48, 196-200 (1982).
4. Thomson, P.D., Harris, N.S.: Detection of plaque-forming cells in the peripheral blood of actively immunized humans. J. Immunol. 118, 1480-1482 (1977).
5. Ochs, H.D., Bohnsack, J., Heller, S.R., Wedgwood, R.J.: Appearance of antibody-producing cells in the peripheral blood in response to immunization with a specific antigen. Birth Defects: Original Article Series 19, 47-50 (1983).
6. Stevens, R.H., Saxon, A.: Immunoregulation in humans: Control of antitetanus toxoid antibody production after booster immunization. J. clin. Invest. 62, 1154-1160 (1978).
7. Lane, H.C., Volkman, D.J., Whalen, G., Fauci, A.S.: *In vitro* antigen-induced, antigen-specific antibody production in man: Specific and polyclonal components, kinetics, and cellular requirements. J. exp. Med. 154, 1043-1057 (1981).
8. Yarchoan, R., Murphy, B.R., Strober, W., Schneider, H.S., Nelson, D.L.: Specific anti-influenza virus antibody production *in vitro* by human peripheral blood mononuclear cells. J. Immunol. 127, 2588-2594 (1981).
9. Zanders, E.D., Smith, C.M., Callard, R.E.: A micro method for the induction and assay of specific *in vitro* antibody responses by human lymphocytes. J. Immunol. Meth. 47, 333-338 (1981).
10. Durandy, A., Fischer, A., Griscelli, C.: Specific *in vitro* anti-mannan antibody production by human blood lymphocytes. Birth Defects: Original Article Series 19, 41-45 (1983).

11. Kelly, B.S., Levy, J.G., Sikova, L.: The use of the enzyme-linked immunosorbent (ELISA) for the detection and quantification of specific antibody from cell cultures. Immunology 37, 45-52 (1979).

12. Kilburn, D.G., Shematek, G., Crane, S.: Quantification of helper factor activity. I. Utilization of the ELISA method to measure helper activity. In: "The Secondary Antibody Response"., J. Immunol. Meth. 44, 301-310 (1981).

13. Boerrigter, G.H., Vos, A., Scheper, R.J.: Direct measurement of in vitro antibody production using an enzyme-linked immunosorbent assay. J. Immunol. Meth. 61, 377-384 (1983).

14. Czerkinsky, C.C., Nilsson, L., Nygren, H., Ouchterlouy, O., Tarkowski, A.: A solid-phase enzyme-linked immunospot (ELISPOT) assay for enumeration of specific antibody-secreting cells. J. Immunol. Meth. 65, 109-121 (1983).

15. Ershler, W.B., Moore, A.L., Hacker, M.P.: Specific in vivo and in vitro antibody response to tetanus toxoid immunization. Clin. Exp. Immunol. 49, 552-558 (1982).

16. Moore, A.L., Ershler, W.B., Hacker, M.P.: Specific antibody synthesis in vitro. I. Technical considerations. J. Immunol. Meth. 70, 13-21 (1984).

17. Ershler, W.B., Moore, A.L., Hacker, M.P., Ninomiya, J., Naylor, P., Goldstein, A.L.: Specific antibody synthesis in vitro. II. Age-associated thymosin enhancement of anti-tetanus antibody synthesis. Immunopharm. 8, 69-77 (1984).

18. Ershler, W.B., Moore, A.L., Socinski, M.A.: Specific antibody synthesis in vitro. III. Correlation of in vivo and in vitro antibody response to influenza immunization in young and old subjects. J. Clin. Lab. Immunol. 16, 63-67 (1985).

19. Ershler, W.B., Moore, A.L., Socinski, M.A.: Influenza and aging: Age-related changes and the effects of thymosin on the antibody response to influenza vaccine. J. Clin. Immunol. 4, 445-454 (1984).

20. Murphy, B.R., Tierney, E.L., Barbour, B.A., Yolken, R.H., Alling, D.W., Holley, P.L., Mayner, R.E., Chanock, R.M.: Use of the enzyme-linked immunosorbent assay to detect serum antibody responses of volunteers who received attenuated influenza A virus vaccines. Infect. Immunol. 29, 342-347 (1980).

21. Volkman, D.J., Lane, H.C., Fauci, A.S.: Antigen-induced in vitro antibody production by humans: A model for B cell activation and immunoregulation. Proc. natn. Acad. Sci. U.S.A. 78, 2528-2531 (1981).

22. Center for Disease Control, Department of Health and Human Sciences. Influenza Vaccines 1982-83. Recommendation of the Immunization Practices Advisory Committee. Ann. intern. Med. 97, 581-583 (1982).

23. Phair, J., Kaufman, C.A., Bjornson, A.: Failure to respond to influenza vaccine in the aged: Correlation with B cell number and function. J. Lab. Clin. Med. 92, 822-828 (1978).

24. Wilcox, H.N.A.: Thymus dependence of the antibody response to tetanus toxoid in mice. Clin. Exp. Immunol. 22, 341-347 (1975).

25. Moore, A.L., Ershler, W.B., Socinski, M.A., Greene, C.J.: Specific anti-influenza antibody synthesis and aging: in vitro effects of thymosin. Viruses, Immunity and Immunedeficiency: International Symposium in the Biomedical Science 1984. Plenum Press, New York, (in press).

DETECTION OF ANTIBODIES TO DNA BY ENZYME-LINKED IMMUNOSORBENT ASSAY (ELISA)

R. Smeenk
Department of Autoimmune Diseases, Central Laboratory of the Netherlands Red Cross Blood Transfusion Service and Laboratory of the Experimental and Clinical Immunology, University of Amsterdam, Amsterdam, The Netherlands

Introduction

Systemic Lupus Erythematosus (SLE) is a disease of unknown aetiology characterized by widespread organ involvement, in particular vasculitis, serositis, and synovitis, as well as renal - central nervous system - and skin disease. Serum from patients with SLE has been found to contain a wide spectrum of autoantibodies, mostly directed against nucleic acids, or proteins that occur closely associated with nucleic acids (1).
Of these autoantibodies, those against deoxyribonucleic acid (DNA) have been studied most extensively, probably because of the easy availability of pure antigen. This has resulted in the development of numerous techniques for the detection, characterization and quantitation of antibodies to DNA. Anti-DNA has been found to be specific for SLE, to a greater or lesser degree depending on the method used, contributing to the clinical usefulness of anti-DNA detection.

In addition to diagnostic use, prognostic benefit of anti-DNA detection has also been advocated; in general, close correlation is found between anti-DNA profile and disease activity in individual patients (2). This is not surprising considering the theory that disease manifestations of SLE are brought about by deposition of immune complexes, composed of anti-DNA and its antigen, formed either in the circulation or in situ (3, 4).

Immunoassay Technology, Vol. 2

Also, differences in antibody characteristics will give rise to differing immune complexes, which is reflected in distinct disease patterns shown by the patients (5, 6).

A completely different view on SLE pathogenesis has recently been advocated by several authors who have described severe cross-reactivity of anti-DNA with phospholipids such as cardiolipin (7, 8), with cell-surface structures (9), with glycosaminoglycans such as heparan sulphate and hyaluronic acid (10), or polynegative structures such as dextran sulphate (11). It has been suggested by some authors (10, 12) that DNA/anti-DNA complexes play no role in the pathogenesis of lupus nephritis, but instead that a negatively charged, cross-reactive antigen is responsible for the tissue damage in SLE (10, 12). Nevertheless, whatever the exact mechanism of tissue injury, the heterogeneity of the anti-DNA population with respect to immunoglobulin (sub) class (13, 14), avidity (15-18) and specificity (19, 20) is bound to play a role. On the basis of this heterogeneity, different reaction patterns of anti-DNA in the various assays are not only to be expected, but have also been found (21-23). The antigen, DNA, is a highly charged polyanion, so assays without an intrinsic check on the immunoglobulin nature of DNA-binding material could be subject to falsely positive results due to non-specific interactions of all sorts of positively charged serum components with the antigen (24, 25).

Assays that specifically detect immunoglobulin binding to DNA, such as the immunofluorescence technique (IFT) on *Crithidia luciliae* (26), will be devoid of unwanted DNA-binding by serum components other than immunoglobulin. However, with the latter technique, anti-DNA cannot be properly quantitated. The Enzyme-Linked Immunosorbent Assay (ELISA) (27) can easily be adapted to the detection of anti-DNA; DNA is coated to a plastic surface, which is then incubated with the serum of a patient after which bound antibodies are detected by means of anti-immunoglobulin antibodies to which an enzyme has been conjugated. The latter reaction is easily visualized by means of exposing the enzyme to a chromogenic or fluorogenic substrate.

In principle, the ELISA combines an inherent check on the immunoglobulin nature of DNA-binding material, a feature shared with the IFT on *Crithidia luciliae*, with the possibility of quantitating the amount of anti-DNA contained in the serum, a feature which is shared with radioimmunoassays like the Farr-assay and PEG-assay (28). The use of enzyme-linked antibodies to detect anti-DNA was first described in 1974 (29). Since then, the method has gained a respected place in many laboratories, and a large number of papers have been devoted to the usefulness of the ELISA in anti-DNA determination. In this chapter, technical details of the method will be summarized, and papers dedicated to the clinical relevance of anti-DNA detection by ELISA will also be discussed.

Technical Aspects of the Anti-DNA ELISA

A. Coating conditions

The first papers on detection of anti-DNA by ELISA described the coating of singlestranded DNA (ssDNA) on to plastic surfaces (tubes or microtitre plates) (29-31). Singlestranded DNA can be coated to plastic directly, by incubating the latter with a dilution of DNA in buffer (e.g. phosphate buffered saline, PBS, pH 7.4). The concentration of DNA used varies according to different authors, but is mostly somewhere in the range 1-30 μg/ml. The optimal concentration probably depends firstly on the brand of plates used, followed by conditions such as buffers, length of incubation period, temperature of incubation etc.

Concerning the coating of doublestranded DNA (dsDNA) to plastic, conflicting results have been published. It was found (32) that direct coating of dsDNA to plastic was virtually impossible, and protamine sulphate was used as an intermediate agent. The results have been confirmed elsewhere (33-37), but in some cases, successful direct coating of dsDNA to plastic has been reported (38-41). Instead of protamine sulphate, other intermediate agents which have been used to precoat ELISA plates are poly-L-lysine (33) and methylated bovine serum albumin (mBSA, 34, 37).

In our hands, in the absence of an intermediate agent, less than 5% of a solution of 2 µg/ml radiolabelled, purely double-stranded, circular PM2-DNA became bound to plastic during a 16 hr incubation period at room temperature (Fig. 1; 42). Nine types of microtitre plates were compared (36) and it was found that the Costar 3590 and Dynatech plates were the most satisfactory with respect to reproducibility of the method. Optimal protamine sulphate concentrations were reported to be 1 mg/ml for the Dynatech plate and 1 µg/ml for the Costar plate, using a precoating incubation period of 90 minutes. We generally precoat plates (96 well micro-ELISA plates from Dynatech) with a solution of 0.5 mg/ml protamine sulphate (Organon, Oss, The Netherlands) in distilled water (150 µl/well), for a period of 2 hr at room temperature (11, 42). After this precoating procedure, plates must be extensively washed, preferably with PBS (3 times) and distilled water (3 times). Use of detergents like Tween in the washing procedure at this time would result in elution of the protamine sulphate from the plastic (42).

DNA from various sources (calf thymus, *Escherichia coli*, rat liver, herring or salmon sperm) has been used as coating antigen. A comparison of these antigens in both native and denatured form was made (35). No significant differences were found between various antigens employed and the use of calf thymus DNA was recommended as this is readily available commercially in an acceptably pure form. Whereas some authors go to the trouble of purifying commercially obtained DNA further, either by protease treatment (32, 35) or by methylated albumin Kieselguhr (MAK) chromatography (38), or to eliminate single-stranded regions in the DNA by S1-nuclease treatment (37), others use the DNA as purchased (31, 40). We tested some 50 sera using plates coated either with commercially available calf thymus DNA (Sigma; 2 µg/ml in PBS) or with highly purified

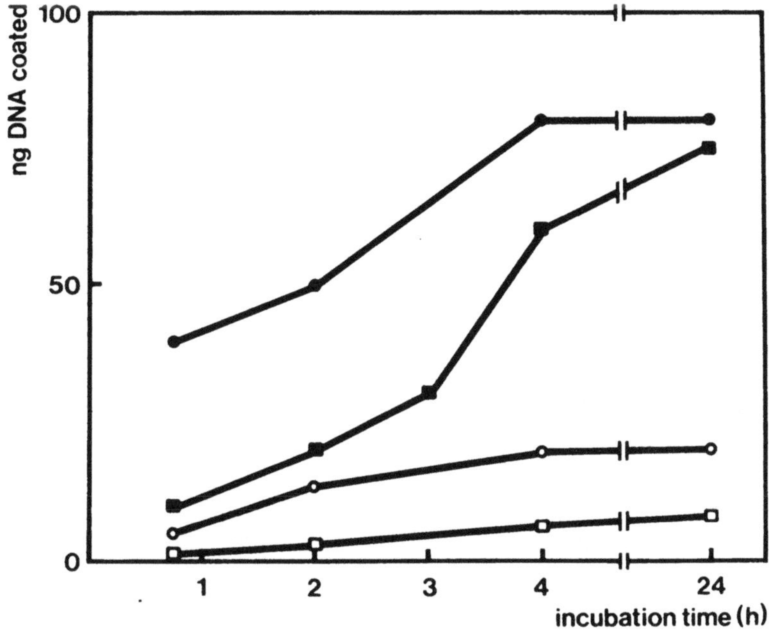

Fig. 1. Kinetics of coating DNA to micro-ELISA plates (42). Two hundred µl of a solution of 1 µg/ml (□/■) or of 5 µg/ml (o/●) ^{3}H-labelled PM2-DNA were incubated for various times with wells precoated (closed symbols) or not precoated (open symbols) with 200 µl protamine sulphate (0.5 mg/ml). The amount of DNA bound to the plastic was calculated from counting 150 µl of the supernatant for remaining radioactivity, and by counting of ^{3}H-DNA eluted from the plastic by means of 2N NaOH. Both methods led to the same results.

purely double stranded circular PM2-DNA (42). From these comparisons we concluded that it made no significant difference whether plates were coated with ds-PM2-DNA, or with calf thymus DNA. Actual data on four of these sera are depicted in Fig. 2.

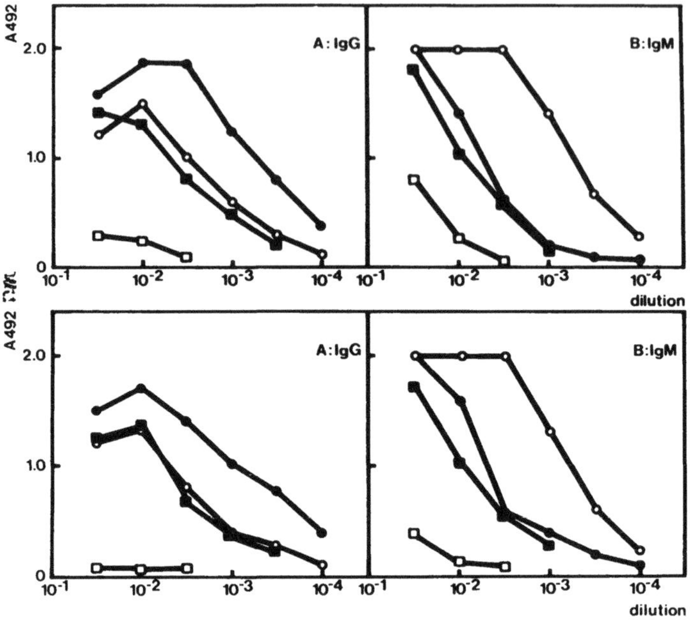

Fig. 2. Comparison of PM2-DNA coated plates (top) with calf thymus-DNA coated plates (bottom) (42). Serial dilutions of serum *Br (●-●), serum *Mo (o-o), serum *Ja (■ - ■) and control serum C1840 (□ - □) were tested in a standard ELISA using either anti-IgG (A) or anti-IgM (B) conjugate. (* Abbreviations of patients' names).

In the Farr-assay, pure doublestrandedness of the DNA is mandatory to make the assay specific. The reason why ELISA puts less strict limitations on the DNA used will be discussed later. Various buffers in which to dilute DNA, as well as different incubation periods have been described. We obtained optimal coating if the DNA was diluted in PBS (2 µg/ml), and incubated (150 µl/well) for 16 hr (overnight) at room temperature with the micro-ELISA plates.

B. Testing of sera

There seems to be very little concordance in the ELISA systems used; many different buffers in which serum dilutions can be made, have been described. Also, to diminish non-specific binding, some authors block free binding places on plastic by - sometimes lengthy - incubations with proteins such as human or bovine serum albumin (38) or gelatine (37), in concentrations between 1 and 20 mg/ml. We have found no need for this blocking coating; after coating with DNA, we washed the plates thoroughly, firstly 3 times with PBS, and then 3 times with distilled water. We diluted the sera to be tested (starting at 1:10) in PBS containing 2% BSA and 0.02% Tween-20 (Baker Chemicals) and incubated the serum dilutions (100 µl/well) with the DNA coated plates for 2 hr at $37^{\circ}C$.

Recently, an attempt has been made to study the influence of assay conditions on anti-DNA ELISA systems, especially with respect to the buffer in which the tested monoclonal antibodies to DNA are diluted (43). The authors often found stronger DNA-binding in Tris-NaCl (50 mM Tris HCl pH 7.5, 100 mM NaCl) than in PBS (137 mM NaCl, 2.7 mM KH_2PO_4, 6 mM Na_2HPO_4, pH 7.5). However, we have found that increased DNA-binding in Tris-buffers, as compared with PBS, tends to follow from non-specific electrostatic interactions of serum proteins such as low density lipoproteins (LDL), complement component Clq and immunoglobulin with DNA (R. Smeenk, unpublished results). As control monoclonals have not been tested (43), we will, for the time being, have to await further confirmation of results.

Following removal of the diluted serum and washing of the plates - 3 times with PBS containing 0.02% Tween (use of Tween at this stage does not lead to elution of DNA or protamine, probably because the interaction with antibodies has stabilized the coating), and 3 times with distilled water - we arrive at the incubation with enzyme-conjugated anti-immunoglobulin antibodies ("conjugate"). Two enzymes are used in this respect: some people prefer alkaline phosphatase conjugated to immunoglobulin (32, 35, 36, 39, 41), whereas others use horse-radish peroxidase (34, 37, 38, 40). In both instances, ammonium sulphate precipitated serum of sheep, goats or rabbits, immunized with (human or murine) immunoglobulins, is used to conjugate with the enzyme.

Various buffers are used for making dilutions of these conjugates. We use the same buffer as the one in which the sera were diluted (PBS containing 2% BSA and 0.02% Tween) for the purpose of diluting the rabbit-anti-human IgG and IgM, both conjugated to horse-radish peroxidase and purchased from Dako Immunoglobulins A/S. Incubation is allowed to take place for 1 hr at 37°C, after which the plates are washed again.

In order to test murine monoclonal anti-DNA, however, we follow a different protocol. Following reaction of (diluted) culture supernatants or ascitic fluid with DNA-coated plates, and the washing of the plates, they are incubated for 1 hr at 37°C with a dilution of goat-anti-mouse immunoglobulin antiserum in PBS, containing 10% normal goat serum and 0.02% Tween-20, and then for 30 min at 37°C with a dilution of monoclonal horse-radish peroxidase (HRP)/anti-HRP complexes (PAP complexes), also made in PBS and containing 10% normal goat serum and 0.02% Tween-20 (11). They are finally washed 3 times with PBS containing 0.02% Tween-20, and 3 times with PBS.

Development of the plates depends in the first place on the conjugate used. Most people who use alkaline phosphatase use p-nitrophenyl-phosphate (Sigma) as a chromogenic substrate to the enzyme. However, concentrations used vary (1-10 mg/ml), as do diluting buffers (either a carbonate buffer plus 1mM $MgCl_2$, pH 9.8 or diethanolamine, pH 9.8 plus 0.5-1 mM $MgCl_2$). Reaction is allowed to take place for 30 min-60 min and is stopped by addition of 10% 2M NaOH. Plates are read at 410 nm.

There is somewhat more variation in chromogenic substrates employed with peroxidase conjugates: 5-aminosalicylic acid (1 mg/ml), ortho phenylene-diamine (OPD; 2 mg/ml), 2,2'-azino-di-(3-ethyl-benzthiazoline-6-sulphonic acid (ABTS, 1 mg/ml), and 3,5,3',5'-tetra-methylbenzidine (TMB, 100 µg/ml) have all been used in anti-DNA ELISA's. From our work, 5-aminosalicylic acid seems to be the less sensitive substrate, whereas we stopped working with OPD because of its known carcinogeneity. We are currently using TMB, diluted in 0.11 M sodium acetate buffer pH 5.5, supplemented with 0.003% H_2O_2. One hundred µl of this is added to each well; after 15 min the reaction is stopped by addition of an equal volume of 2 M H_2SO_4 to the wells. Plates are then read 450 nm in a Titertek Multiskan.

As an alternative to chromogenic substrates, the use of fluorogenic substrates with peroxidase conjugates (40) and with alkaline phosphatase conjugates (44) has also been described. Substrates which have been used are p-hydroxyphenylacetic acid and 4-methyl-umbelliferyl phosphate, respectively. Authors using these fluorogenic substrates claim a higher sensitivity of the technique compared with assays using chromogenic substrates. In our hands, reproducibility of all ELISA's has not been as good as is often claimed; the units or titres obtained may vary from day to day as much as two-fold (42).

Specificity in SLE of Anti-DNA Detected by ELISA

To evaluate the specificity of anti-DNA in SLE, it is probably necessary to differentiate between anti-dsDNA and anti-ssDNA. Conflicting reports have been published in the past on the significance of antibodies to dsDNA and to ssDNA. Use of ssDNA as antigen, for instance in the Farr-assay, has often been said to lead to loss of specificity for SLE (45). However, this may not only be due to the occurrence of anti-ssDNA in non-SLE patients, but mainly to the fact that serum proteins will bind to ssDNA much better (i.e. with a much higher avidity) than to dsDNA (23).

Methods such as the Farr-assay, which do not incorporate a check on the immunoglobulin nature of DNA-binding material will, therefore, be subject to falsely positive results, due to non-specific (i.e. non-Ig) ssDNA binding. Assays such as the ELISA technique possess an inherent check on the Ig-nature of DNA-binding substances, which overcomes such problems. It is our opinion that assays incorporating such inherent checks have a lesser need for pure double-strandedness of the DNA used as antigen, although the binding of anti-DNA with ssDNA will be of a higher avidity than with dsDNA, as a result of the greater flexibility of the ssDNA antigen (23, 46). By using ssDNA as antigen in contrast to dsDNA, this will generally lead to higher titres being obtained.

There may, therefore, be populations of purely anti-dsDNA and anti-ssDNA antibodies, although the majority of anti-DNA antibodies probably react with both antigens, but perhaps with differing avidities (21). On the other hand, high avidity anti-DNA has been found more specific to SLE than low avidity anti-DNA (18, 47), which could be a reason to stick to the use of purely doublestranded DNA in the anti-DNA ELISA.

Data on the specificity of anti-DNA ELISA, using either dsDNA or ssDNA, have been compiled from the literature, in Table 1. A general tendency that can be deduced from this table is that normal sera are hardly ever found to be positive, using either ssDNA or dsDNA as antigen. Taking into account the rather divergent technical procedures used by the various authors, there is, nevertheless, much agreement with respect to the percentage of SLE patients found to be positive. It is also apparent that anti-ssDNA is more often found in rheumatoid arthritis (R.A.) than is anti-dsDNA, the antigen used (48) probably containing a large proportion of ssDNA.

In Table 2, data from testing 551 sera of normal human controls and patients with well-defined diseases are given (42). It can be seen from Table 2 and Fig. 3, where data on these sera have been depicted quantitatively, that IgM anti-DNA was seldom detected in sera of patients with autoimmune disorders other that SLE, with the exception of myasthenia gravis. IgG anti-DNA was found more often in these sera, especially in sera of patients with autoimmune thyroiditis, Sjögren's syndrome, scleroderma, myasthenia gravis and autoimmune gastritis.

Table 1. Specificity of anti-DNA detected by ELISA to SLE

Antigen	Author (ref.)	Percentage of sera with detectable anti-DNA					
		NHS	SLE	RA	SS	Sclero-derma	A.I. liver disease
(a) ssDNA	Jones (40)	2.4	90	20	-	-	-
	Klotz (32)	3.0	80	*	-	-	-
	Stokes (35)	0.0	100	50	-	-	25
(b) dsDNA	Jones (40)	0.0	60	0	-	-	-
	Klotz (32)	0.0	53	-	-	-	-
	Stokes (35)	0.0	100	5	-	-	25
	Karsh (48)	3.3	85	26	47	-	-
	Eaton (36)	2.5	80	7	5	5	-
	Smeenk (42)	0.0	100	2	21	33	14
	Rubin (37)	5.0	25	0	0	16	-

NHS: Normal human serum; SLE: systemic lupus erythematosus; RA: rheumatoid arthritis; SS: Sjögren's syndrome; A.I.: autoimmune.
* Not tested.

Table 2. ELISA for anti-DNA in different patient groups (42)

Patient group	Number of sera tested	Positive for anti-DNA	
		IgG	IgM
SLE	75	75	66
Rheumatoid arthritis	63	2	1
Thyroid autoimmune disease	66	9	0
Sjögren's syndrome	1	3	0
Scleroderma	12	4	0
Addison's disease	33	0	0
Autoimmune haemolytic anaemia	39	3	0
Autoimmune gastritis	45	8	0
Myasthenia gravis	25	5	4
Autoimmune liver disease	61	9	1
Normal control sera	118	0	0

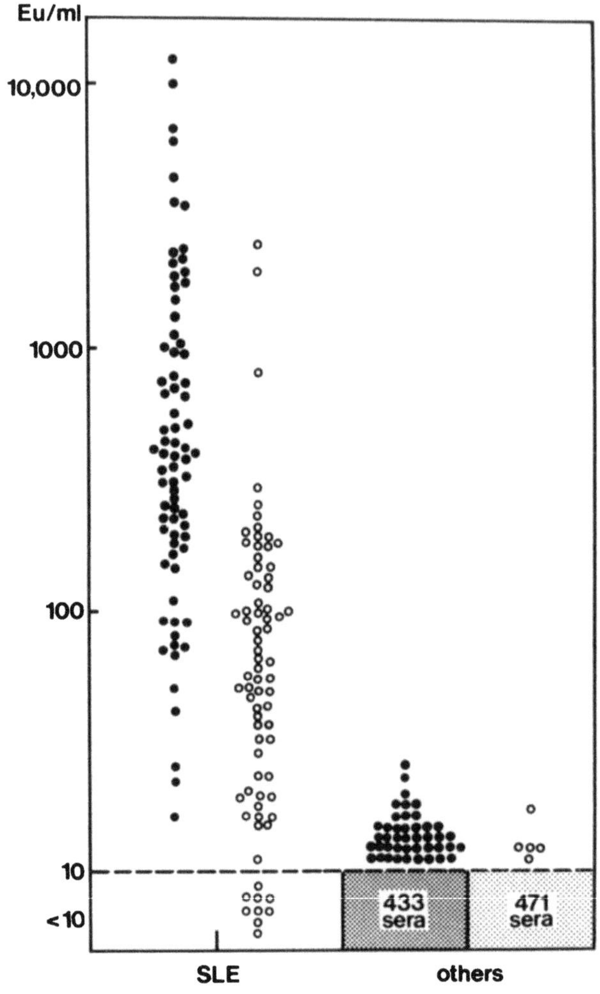

Fig. 3. ELISA for anti-DNA in different patient groups, using anti-IgG (●-●) or anti-IgM (o-o) conjugate (42).

However, in general these sera contained relatively little anti-DNA, i.e. less than 20 EU/ml; a serum is said to contain 1 EU/ml of anti-DNA if 100 µl of undiluted serum would give an absorption of 1, screened at a 1:10 dilution and sera containing more than 10 EU/ml are considered positive. In comparison, the mean IgG anti-DNA contents of 75 sera from SLE patients was 1142 EU/ml.

In conclusion, the anti-DNA ELISA is rather specific for SLE, although low levels of anti-DNA may also be detected in several other autoimmune diseases.

Comparison of the Anti-DNA ELISA with other Anti-DNA Assays

Several authors have compared the anti-DNA ELISA with other commonly used anti-DNA assays such as the IFT on *Crithidia luciliae*, Farr-assay, PEG-assay and cellulose acetate or millipore filtration assay. Generally, these methods were found to be in good agreement with the anti-DNA ELISA (Table 3).

Table 3. Correlation of anti-DNA ELISA with the IFT on *Crithidia luciliae* and the Farr-assay.

		Coefficient of correlation between ELISA and:					
Author (ref)	Number of SLE sera tested	IFT on *Crith. luc.*			Farr-assay		
		IgM		IgG	IgM		IgG
Eaton (36)	196		0.67			-	
Smeenk (42)	75	0.91		0.73		0.67	
Feldman (41)	59	-		0.80		-	
Stokes (35)	98		-			0.62	
Rubin (37)	16		-			0.96	
Miller (38)	88		-			0.65	
Karsh (48)	96		-		0.55		0.72

It was found (36) that, of 154 sera of SLE patients, 91 were positive in both the anti-DNA ELISA and the IFT on *Crithidia luciliae*, whereas 11 sera were negative in both assays. Some discrepancies were observed too because 48 sera were ELISA-positive and IFT-negative. The coefficient of correlation between ELISA and IFT on *Crithidia luciliae* for these 154 sera was still rather high($r=0.674$ and $p<0.001$). The anti-DNA ELISA was also compared with the IFT on *Crithidia luciliae* (40). Results obtained showed discrepancies in 13 of the 42 sera tested. Seven of these were ELISA-positive but IFT-negative, and 6 were ELISA-negative but IFT-positive. The correlation between anti-DNA ELISA and GF/A filter assay was studied (39), when a general, but not absolute, correlation between both assays was observed. Several groups of workers compared the anti-DNA ELISA with the Farr-assay (35, 37, 38, 40, 41, 42, 48) and, generally, good correlations were obtained, though discrepancies were also noticed. It was found (40) that 4 sera (out of 48 sera tested) were ELISA-positive but Farr-negative and that 8 sera were Farr-positive but ELISA-negative. It was observed (35) that a rather high percentage (14%) of the 98 sera from SLE patients tested, were ELISA-positive but Farr-negative, yet the calculated overall correlation between the Farr-assay and the anti-DNA ELISA was 0.62 ($p<0.001$).

Separate correlations were calculated (48) for IgG-anti-DNA and IgM-anti-DNA, measured by ELISA, with the Farr-assay, and a somewhat higher correlation between the IgG-anti-DNA ELISA and the Farr-assay ($r=0.62$) than between the IgM-anti-DNA ELISA and the Farr-assay ($r=0.55$) was obtained. A very high correlation between the anti-DNA ELISA and the Farr-assay ($r=0.96$) was found (37), when it was claimed that anti-histone antibodies may react with protamine sulphate which is used as a precoating agent. According to the findings, the use of mBSA instead of protamine sulphate as a precoating agent, increases the specificity of the ELISA for antibodies to DNA. It was stated (38) that the anti-DNA ELISA has a sensitivity and specificity comparable to the Farr-assay for patients with active disease (with or without renal involvement), but that the ELISA detected anti-DNA more frequently in patients with inactive disease.

The correlation between IgG-anti-DNA detected by ELISA and the Farr-assay has also been calculated (41) and a high correlation ($r=0.80$) for the 59 sera of SLE patients tested was obtained. We compared the anti-DNA ELISA with the Farr-assay, the PEG-assay and the IFT on *Crithidia luciliae*, using sera of 75 patients with SLE (42). We, too, obtained high correlations between the ELISA and the Farr-assay (r=0.67) and between the ELISA and the PEG-assay (r=0.69). As for the IFT on *Crithidia luciliae* and the anti-DNA ELISA, separate correlations were calculated for IgG-anti-DNA (r=0.73) and IgM-anti-DNA (r=0.91). Notwithstanding the rather high coefficients of correlation obtained, discrepancies were also observed. An example can be found in Table 4, where results of 4 sera from SLE patients obtained with different anti-DNA assays are depicted.

Table 4. DNA-binding of 4 sera, measured with 4 different techniques (42).

Serum	ELISA		IFT		Farr-assay	PEG-assay
	IgG	IgM	IgG	IgM	(U/ml)	(U/ml)
Br	1200*	250*	1/2560+	1/80+	4800§	570§
Mo	200	2000	1/640	1/640	450	270
Ja	200	100	1/1280	1/20	< 10	250
C1840	< 20	< 20	< 1/10	< 1/10	< 10	< 10

* DNA-binding expressed as EU/ml (reciprocal dilution giving an absorbance of 1.0).
+ DNA-binding expressed as a titre (last dilution giving a positive fluorescence).
§ DNA-binding expressed as U/ml (1 unit is defined as the amount of anti-DNA precipitating 30% of 100 ng (^{3}H)-PM2 DNA in a standardized Farr-assay or PEG-assay (28)).

As can be deduced from Table 4, anti-dsDNA of serum Ja cannot be detected by the Farr-assay, from which we conclude that it is of low avidity (18, 22). As the ELISA detects anti-dsDNA in the serum, it can be concluded that the ELISA can also measure antibodies to dsDNA of low avidity.

It has been reported that rising anti-DNA levels, under

defined conditions, always predict an exacerbation of SLE, whereas the exacerbation itself coincides with a sharp drop in anti-dsDNA level (2). Using the ELISA, we examined a series of sera from three different patients in the period that they showed exacerbation of their disease. The patients were selected on the basis of different behaviour of their anti-DNA antibodies in the Farr-assay and the PEG-assay. Serum of patient Bu (Fig. 4A) was both Farr- and PEG-assay positive at the time of exacerbation, serum of patient Ba (Fig. 4B) was Farr-positive but PEG-negative, and serum of patient SB (Fig. 4C) was PEG-positive but Farr-negative. It is clear that anti-dsDNA in all of these patients was readily detected by means of the ELISA with anti-IgG conjugate. Peaks in anti-DNA levels measured this way, mostly coincided with peaks in the Farr-assay and/or the PEG-assay. In 2 of the 3 patients, the ELISA even detected anti-DNA at times before (and after) the exacerbation, where both radioimmunoassays did not detect anti-DNA. Using anti-IgM conjugate, these patterns differed; serum of patient Ba apparently does not have (detectable) IgM-anti-DNA; in the serum of patient Bu, anti-DNA of the IgM-class more or less follows the IgG-anti-DNA pattern; but in the serum of patient SB, IgM anti-DNA seems to precede IgG anti-DNA. This IgM anti-DNA seems not to be detected by the Farr- or PEG-assays.

On the Avidity of Anti-DNA Detection by ELISA

From the previous section, it should have become clear that the ELISA for anti-DNA is capable of detecting anti-DNA of low avidity, which is missed by the Farr-assay. The method shares this feature with the IFT on *Crithidia luciliae* and the PEG-assay (18, 22, 42). Yet, the method even seems to detect antibodies missed by the latter two assays. The question arises as to whether these antibodies are true anti-DNA antibodies. As the interaction between anti-DNA and DNA has been found to be purely electrostatical (49, 50), the possibility exists that positi-

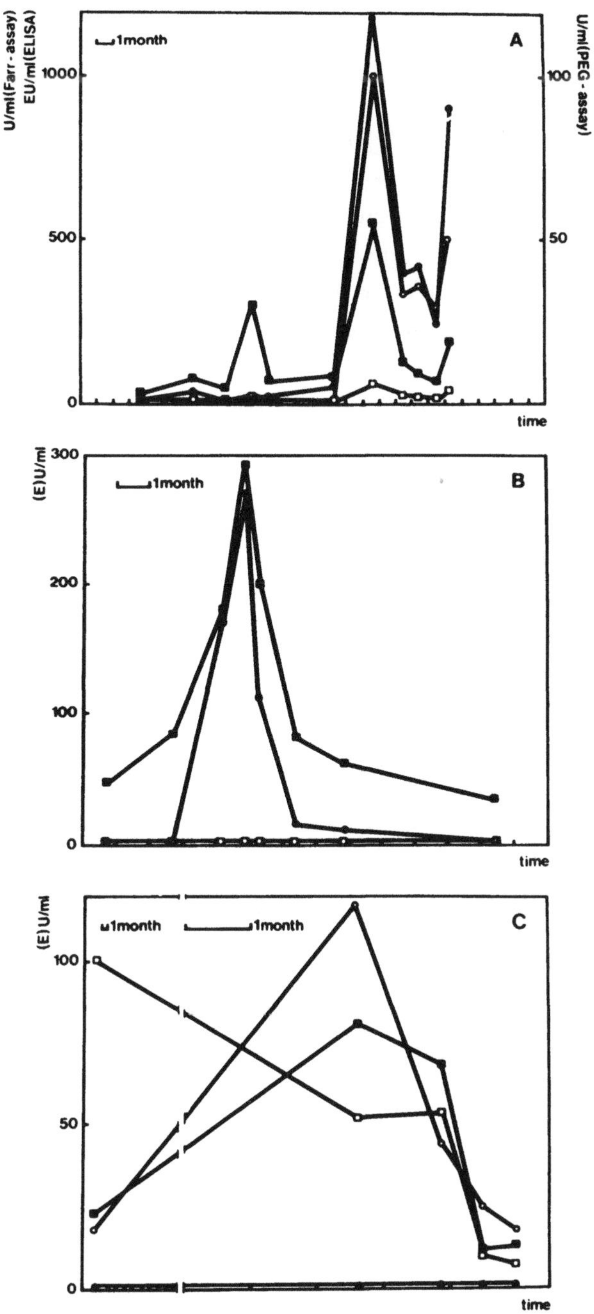

Fig. 4. Longitudinal patient studies (42). Serum anti-DNA levels were measured by means of the Farr-assay(●-●), the PEG-assay (o-o) and the ELISA using anti-IgG (■ - ■) or anti-IgM (□ - □) conjugate. (A) patient Bu; (B) patient Ba; (C) patient SB.

vely charged immunoglobulins, that are not really antibodies to DNA, bind to DNA. Such anti-polynegative antibodies might form the basis of the frequently observed cross-reactions of monoclonal antibodies to DNA (7-9) with various negatively charged molecules (11).

An important point discussed (51) deals with the question of whether ELISAs measure antibody concentration or antibody affinity/avidity. From this review (51) it becomes clear that the ELISA is indeed a sensitive technique for the measurement of antibodies of high affinity, with a detection limit of about 1 ng/ml of antibody. However, for low affinity antibodies, the detection limit may be no less than 2 µg/ml. Of course, these considerations have strong implications for the interpretation of Optical Density (O.D.) values in the ELISA. Whether end-point titres (maximum serum dilution giving the standard O.D.-reading above control) or single-dilution O.D.-readings (which are then related to a standard serum) are used, it would seem impossible to quantify the antibody levels in an ELISA without making assumptions about the affinity of the antibodies.

Taking this into account, the ELISA should primarily be used as a qualitative, or at most as a semi-quantitative assay. Also, the density at which antigen is coated to the plates seems to influence assay results. It was found (52), in a BSA/anti-BSA system, that samples containing large amounts of low avidity antibodies gave higher absorbance values when plates of lower antigen density were used.

Concluding Remarks

In the past few years, the ELISA for antibodies to DNA has been proclaimed as a sensitive assay, giving quantitative results, comparable to other anti-DNA assays. In part, this is certainly true. The assay detects antibodies of low avidity, which are missed by the Farr-assay and circumvents the necessity to work with radiolabelled antigens as used in both the Farr-assay and the PEG-assay. Furthermore, the ELISA incorporates an inherent check on the immunoglobulin nature of what binds to DNA, as does the IFT on *Crithidia luciliae*. Yet, both from the observed differences in detecting antibodies of different affinities (51, 52) and from day-to-day reproducibility (42), restrictions arise as to use of the anti-DNA ELISA as a truly quantitative assay.

If used diagnostically, one should keep in mind that anti-DNA of low avidity (which is co-detected with this assay) has less specificity to SLE than anti-dsDNA of high avidity (detected by the Farr-assay). On the other hand, information on the avidity of the anti-DNA may be valuable to the clinician. We found that anti-DNA of high avidity is significantly correlated to the occurrence of nephritis in SLE, whereas anti-DNA of low avidity is more often found in patients with central nervous system involvement (42, 53).

If the anti-DNA ELISA is to be used to quantitate amounts of antibody present in patients' sera, the shortcomings of the method should be kept in mind, and the assay should preferably be used in a semi-quantitative way. To facilitate comparison of results between different institutes, an internationally accepted reference serum, to which titres or units are recalculated, should be incorporated in each study.

Acknowledgements

The author wishes to thank Trees Duin and Omar Suarez for their technical help in performing these studies, Tom Swaak for sera and clinical information on the patients we studied and Lucien Aarden for fruitful discussions on the subject. Lucien Aarden, Ineke v/d Schaal and Toon Westgeest are also thanked for critically reading the manuscript, and special credits go to Yvonne Kluyt and Annette Lochtenberg for their expert skill and speed in preparing the typescript.

References

1. Tan, E.M.: Autoantibodies to nuclear antigens (ANA): Their immunobiology and medicine. In: "Advances in Immunology", vol. 33, Eds. Dixon, F.J., Kunkel, H.G., Academic Press, New York, pp. 167-241 (1982).

2. Swaak, A.J.G., Aarden, L.A., Statius van Eps, L.W., Feltkamp, T.E.W.: Anti-dsDNA and complement profiles as prognostic guides in SLE. Arthr. Rheum. 22, 226-235 (1979).

3. Koffler, D., Agnello, V., Thoburn, R., Kunkel, H.G.: SLE: prototype of immune complex nephritis in man. J. exp. Med. 134, 169S-179S (1971).

4. Winfield, J.B., Koffler, D., Kunkel, H.G.: Role of DNA/anti-DNA complexes in the immunopathogenesis of tissue injury in SLE. Scand. J. Rheumatol. suppl. 11, 59-62 (1975).

5. Griffiths, I.D., Holian, J., Maini, R.N., Glass, D.N., Scott, J.T.: Anti-DNA antibodies: quantitative and qualitative aspects of measurement using ^{125}I-labelled DNA. Ann. rheum. Dis. 36 (suppl.) 122-128 (1977).

6. Ballou, S.P., Kushner, I.: Immunochemical characteristics of antibodies to DNA in patients with active SLE. Clin. exp. Immunol. 37, 58-67 (1979).

7. Lafer, M., Rauch, J., Andrzejewski, C. Jr., Mudd, D., Furie, R., Furie, B., Schwartz, R.S., Stollar, B.D.: Polyspecific monoclonal autoantibodies reactive with both polynucleotides and phospholipids. J. exp. Med. 153, 897-909 (1981).

8. Shoenfeld, Y., Rauch, J., Massicotte, M., Datta, S.K., Andre-Schwartz, J., Stollar, B.D., Schwartz, R.S.: Polyspecificity of monoclonal lupus antibodies produced by human-human hybridomas. New Engl. J. Med. 308, 414-420 (1983).

9. Jacob. L., Tron, F., Bach, J.F., Louvard, D.: A monoclonal anti-DNA antibody also binds to cell-surface protein(s). Proc. natn. Acad. Sci. U.S.A. 81, 3843-3845 (1984).

10. Faaber, P., Capel, P.J.A., Rijke, G.P.M., Vierwinden, G., Van der Putte, L.B.A., Koene, R.A.P.: Crossreactivity of anti-DNA antibodies with proteoglycans. Clin. exp. Immunol. 55, 502-508 (1984).

11. Smeenk, R., Westgeest, A., Duin, T., Van den Brink, H.: Crossreactivity of monoclonal antibodies to DNA: related to avidity? Protides Biol. Fluids vol. XXXIII (1985) (in press).

12. Eilat, D., Cross-reactions of anti-DNA antibodies and the central dogma of lupus nephritis. Immunol. Today 6, 123-127 (1985).

13. Pennebaker, J.B., Gilliam, J.N., Ziff, M.: Immunoglobulin classes of DNA-binding activity in serum and skin in SLE. J. clin. Invest. 60, 1331-1338 (1977).

14. Sontheimer, R.D., Gilliam, J.N.: DNA antibody class, subclass and complement fixation in SLE with and without nephritis. Clin. Immunol. Immunopathol. 10, 495-497 (1978).

15. Steward, M.W., Glass, D.N., Maini, R.N., Scott, J.T.: Role of low avidity antibody to native DNA in human and murine lupus syndrome. J. Rheumatol. 1, 75-76 (1974).

16. Winfield, J.B., Faiferman, I., Koffler, D.: Avidity of anti-DNA antibodies in serum and IgG glomerular eluates from patients with SLE. J. clin. Invest. 59, 90-96 (1977).

17. Leon, S.A., Green, A., Ehrlich, G.E., Poland, M., Shapiro, B.P.: Avidity of antibodies to DNA: relation to severity of renal involvement. Arthr. Rheum. 20, 23-29 (1977).

18. Smeenk, R., Van der Lelij, G., Aarden, L.: Measurement of low avidity anti-dsDNA by the *Crithidia luciliae* test and the PEG-assay. Clin. exp. Immunol. 50, 603-610 (1982).

19. Papalian, M., Lafer, E., Wong, R., Stollar, B.D.: Reaction of SLE anti-native DNA antibodies with native DNA fragments from 20-1200 base pairs. J. clin. Invest. 65, 469-477 (1980).

20. Ebling, F., Hahn, B.H.: Restricted subpopulations of DNA antibodies in kidneys of mice with systemic lupus. Comparison of antibodies in serum and renal eluates. Arthr. Rheum. 23, 392-403 (1980).

21. Maini, R.N., Holborrow, E.J.: Eds. "Detection and Measurement of Circulating Soluble Antigen-Antibody Complexes and Anti-DNA Antibodies". Ann. Rheum. Dis. (suppl.) 36, 67-81 (1977).

22. Smeenk, R., Van der Lelij, G., Aarden, L.: Avidity of antibodies to dsDNA: Comparison of IFT on *Crithidia luciliae*, Farr-assay and PEG-assay. J. Immunol. 128, 73-78 (1982).

23. Aarden, L., Smeenk, R.: Immunochemical properties of antibodies to DNA and their influence on detection methods. In: "Antibodies to Nuclear Antigens", Eds. Kalden, J.R., Fèltkamp, T.E.W., Excerpta Medica, Amsterdam, pp. 23-28 (1982).

24. Smeenk, R., Aarden, L.: The use of polyethylene glycol precipitation to detect antibodies to dsDNA in SLE. J. Immunol. Meth. 39, 165-180 (1980).

25. Winfield, J.B., Shaw, M., Taylor, R.P., Eisenberg, R.A.: Nature of dsDNA binding in seropositive RA: formation of low avidity DNA/rheumatoid factor/IgG/low-density-lipoprotein complexes. J. Immunol. 126, 1596-1602 (1981).

26. Aarden, L.A., De Groot, E.R., Feltkamp, T.E.W.: Immunology of DNA. III. *Crithidia luciliae*, a simple substrate for the determination of anti-dsDNA with the IFT. Ann. N.Y. Acad. Sci. 254, 505-515 (1975).

27. Engvall, E., Perlmann, P.: Enzyme-linked immunosorbent assay (ELISA). III. Quantitation of specific antibodies

by enzyme-labelled anti-immunoglobulin in antigen-coated tubes. J. Immunol. 109, 129-135 (1972).

28. Aarden, L., Smeenk, R.: Measurement of antibodies specific for DNA. In: "Immunological Methods", vol. II., Eds. Lefkovits, I., Pernis, B., Academic Press, New York, pp. 75-82 (1981).

29. Pesce, A.J., Mendoza, N., Boreisha, I., Gaizutis, M.A., Pollack, V.E.: Use of enzyme-linked antibodies to measure serum anti-DNA antibodies in SLE. Clin. Chem. 20, 353-359 (1974).

30. Engvall, E.: Determination of antibodies to DNA by ELISA. Lancet ii, 1410-1416 (1976).

31. Gripenberg, M., Linder, E., Kurki, P., Engvall, E.: A solid-phase enzyme-linked immunosorbent assay (ELISA) for the demonstration of antibodies against denatured single-stranded DNA in patients' sera. Scand. J. Immunol. 7, 151-159 (1978).

32. Klotz, J.L., Minami, R.M., Teplitz, R.L.: An enzyme-linked immunosorbent assay for antibodies to native and denatured DNA. J. Immunol. Meth. 29, 155-165 (1979).

33. Field, A.K., Davies, M.E., Tynell, A.A.: Determination of antibodies to dsRNA by ELISA. Proc. Soc. exp. Biol. Med. 164, 524-529 (1980).

34. Kavai, M., Banyai, A., Zsindely, A., Sonkoly, I., Szegedi, G.: Enzyme-linked immunosorbent assay for antibodies to native DNA in sera of patients with SLE. J. Immunol. Meth. 48, 169-175 (1982).

35. Stokes, R.P., Cordwell, A., Thompson, R.A.: A simple, rapid ELISA method for the detection of DNA antibodies. J. clin. Path. 35, 566-573 (1982).

36. Eaton, R.B., Schneider, G., Schur, P.H.: Enzyme immunoassay for antibodies to native DNA. Arthr. Rheum. 26, 52-62 (1983).

37. Rubin, R.L., Joslin, F.G., Tan, E.M.: An improved ELISA for anti-native DNA by elimination of interference by anti-histone antibodies. J. Immunol. Meth. 63, 359-366 (1983).

38. Miller, T.E., Lahita, R.G., Zarro, V.J., MacWilliam, J., Koffler, D.: Clinical significance of anti-double stranded DNA antibodies detected by a solid phase enzyme immunoassay. Arthr. Rheum. 24, 602-610 (1981).

39. Pisetsky, D.S., Peters, D.V.: A simple enzyme-linked immunosorbent assay for antibodies to native DNA. J. Immunol. Meth. 41, 187-200 (1981).

40. Jones, C.E., Findlay Pike, J., Dickinson, R.P., Rousseau, R.J.: Fluorescent enzyme immunoassay for antibody to single- or double-stranded DNA. Am. J. clin. Path. 75, 509-518 (1981).

41. Feldman, M.D., Huston, D.P., Karsh, J., Balow, J.E., Klima, E., Steinberg, A.D.: Correlation of serum IgG, IgM and anti-

native DNA antibodies with renal and clinical indexes of activity in SLE. J. Rheumatol. 9, 52-58 (1982).

42. Smeenk, R.: Low avidity antibodies to dsDNA. Thesis, Rodopi, Amsterdam, pp. 141-162 (1982).

43. Pisetsky, D.S., Semper, K.F.: Influence of assay conditions on ELISA determinations of anti-DNA antibodies. J. Immunol. Meth. 74, 217-227 (1984).

44. Ali, A., Ali, R.: Enzyme-linked immunosorbent assay for anti-DNA antibodies using fluorogenic and colorigenic substrates. J. Immunol. Meth. 56, 341-346 (1983).

45. Locker, J.D., Medof, M.E., Bennett, R.M., Sukhupunyaraksa, S.: Characterization of DNA used to assay sera for anti-DNA antibodies: determination of the specificities of anti-DNA antibodies in SLE and non-SLE rheumatic disease states. J. Immunol. 118, 694-701 (1977).

46. Winfield, J.B., Faiferman, J., Koffler, D.: Avidity of anti-DNA antibodies in serum and IgG glomerular eluates from patients with SLE. Association of high avidity anti-native DNA antibody with glomerulonephritis. J. clin. Invest. 59, 90-96 (1977).

47. Smeenk, R., Van der Lelij, G., Swaak, T., Groenwold, J., Aarden, L.: Specificity to SLE of antibodies to dsDNA measured with the polyethylene glycol precipitation assay. Arthr. Rheum. 25, 631-638 (1982).

48. Karsh, J., Halbert, S.P., Anken, M., Klima, E., Steinberg, A.D.: Anti-DNA, anti-deoxyribonucleoprotein and rheumatoid factor measured by ELISA in patients with SLE, Sjögren's Syndrome and Rheumatoid Arthritis. Int. Archs. Allergy appl. Immunol. 68, 60-69 (1982).

49. De Groot, E.R., Lamers, M.C., Aarden, L.A., Smeenk, R.J.T., Van Oss, C.J.: Dissociation of DNA/anti-DNA complexes at high pH. Immunol. Comm. 9, 515-528 (1980).

50. Smeenk, R.J.T., Aarden, L.A., Van Oss, C.J.: Comparison between dissociation and inhibition of association of DNA/anti-DNA complexes. Immunol. Comm. 12, 177-188 (1983).

51. Steward, M.W., Lew, A.M.: The importance of antibody affinity in the performance of immunoassays for antibody. J. Immunol. Meth. 78, 173-190 (1985).

52. Lehtonen, O.-P., Viljanen, M.K.: Antigen density in ELISA: effect on avidity dependency. J. Immunol. Meth. 36, 63-70 (1980).

53. Swaak, A.J.G., Smeenk, R.: Clinical differences between SLE patients in relation to the avidity of their anti-dsDNA. Protides of the Biol. Fluids, vol. XXXIII (1985) (in press).

A COMPARISON OF FOUR DIFFERENT ANTI-DNA ASSAYS

R. Smeenk
Department of Autoimmune Diseases, Central Laboratory of the Netherlands Red Cross Blood Transfusion Service and Laboratory for Experimental and Clinical Immunology, University of Amsterdam, Amsterdam, The Netherlands

Introduction

Systemic Lupus Erythematosus (SLE) is a disease of unknown etiology, characterized by a multitude of different disease phenomena, such as vasculitis, serositis, synovitis, arthritis and nephritis (1). Patients with SLE may present with one or more of these features. Since the first comprehensive description of SLE (2), there has been considerable progress in delineating the immunological defects and pathogenic mechanisms of this disease. SLE has been termed an autoimmune disease because many antibodies directed to autoantigens, in particular nucleic acids and proteins associated with nucleic acids, can be found in the circulation of such patients (3).

Although the term autoimmune disease suggests that disease phenomena are caused by interaction of autoantibodies with antigens present in the patient's own tissue, it is not clear whether this indeed takes place. SLE has often been considered primarily to be an immune complex disease; the antibodies, chronically produced by the patient, combine with antigen to form complexes that actually mediate tissue damage (4-7). It is not known whether these antigens are foreign (perhaps viral) or the patient's own. More recently, a new hypothesis has been formulated by authors who observed profound cross-reactions between antibodies to deoxyribonucleic acid (DNA) and antigens such as phospholipids, cell-membrane constituents, and proteoglycans (8-11). It has been

Immunoassay Technology, Vol. 2

suggested that the pathogenesis of SLE may be mediated through the direct binding of anti-DNA with such negatively charged structures, resulting in profound tissue damage (11, 12).

Since the first description, almost 30 years ago (13-16), of components present in SLE sera reactive with DNA, numerous techniques have been developed for the detection, characterization and quantitation of anti-DNA (17, 18). Early methods showed that the presence of antibodies to DNA was strongly associated with SLE, which stressed the diagnostic value of anti-DNA detection (19). However, when more sensitive methods were developed, antibodies to DNA were also found to occur in other clinical syndromes (20-23). The specificity for SLE of anti-DNA was regained when reaction conditions were more carefully controlled and purely doublestranded DNA (dsDNA) was used as a substrate (24-26). A comparative study, supported by the Arthritis and Rheumatism Council (18), supported the notion that the most valuable sources of dsDNA were PM2 bacteriophages, radiolabelled in vivo, and *Crithidia luciliae*, haemoflagelates containing a giant mitochondrion composed of dsDNA. The former DNA was used in a radioimmunoassay, the Farr-assay (27-29), the latter in an immunofluorescence technique (30, 31). The specificity for SLE of anti-DNA measured by these two methods, as well as the mutual correlation between the two, has been repeatedly confirmed (32-37).

Although good correlation was found between the Farr-assay and IFT on *Crithidia luciliae*, when sera from patients with definite SLE were tested, this correlation was lost when we tested sera referred to our institute for routine anti-dsDNA determination. Less than 50% of the Crithidia-positive sera were found to be Farr-positive (38). Using a radioimmunoassay developed by Riley and co-workers (39, 40) and modified by ourselves (41), we were able to prove that this discrepancy was caused by the fact that DNA/anti-DNA interactions of low-avidity were not detected in the Farr-assay (42, 43). The modified assay, the PEG-assay, uses polyethylene glycol (PEG) to precipitate the DNA/anti-DNA complexes, whereas in the Farr-assay, they are precipitated by ammonium sulphate.

Apart from the IFT on *Crithidia luciliae* and both the above mentioned radioimmunoassays (the Farr-assay and the PEG-assay), the enzyme-linked immunosorbent assay (ELISA) is nowadays frequently used for the detection of antibodies to DNA (44). In this chapter, technical details of the four mentioned anti-DNA assays and their mutual correlations will be discussed.

Technical Details of the Methods

(A) The Farr-assay

This technique, initially introduced for anti-DNA detection (27), has been studied quite extensively (17, 18, 28, 29, 31, 36-40, 45-47). The method has been shown to be very specific for SLE, provided purely doublestranded DNA is used as antigen (18, 29). Radiolabelled circular dsDNA from the bacteriophage PM2 probably serves this aim best (18, 31). It is prepared by culturing PM2 bacteriophages in the presence of (^{3}H)-thymidine, on their host, the bacterium *Pseudomonas* strain BAL41 (preferably a thymidine auxotrophic mutant) (31). Nevertheless, many people use iodine-labelled (calf thymus) DNA, because that can be obtained commercially. This certainly makes the assay more prone to non-specific DNA-binding, as was demonstrated (18, 49). Also, it is more difficult to quantitate results, as the labelled iodine will introduce doublestrand nicks in the DNA, and the size of the DNA influences results (29). The use of different buffers will also influence the assay (46).

We perform the Farr-assay as follows: 50 µl of (a dilution of) serum, 50 µl of a solution of 16 mg/ml human gammaglobulin (HGG; Cohn Fraction II), 50 µl of a solution of 2 µg/ml (^{3}H)-PM2-DNA and 50 µl of phosphate buffered saline (PBS; 0.14 M NaCl, 0.01 M phosphate, pH 7.4) are mixed together and incubated for 1 hr at 37°C. All dilutions are made with PBS. Five ml of a cold, 50% saturated solution of ammonium sulphate (in PBS; pH 7.4) are added, and the precipitate is allowed to form for 30' at 4°C. After centrifugation for 15 min at 3000 g in a refrigerated centrifuge (4°C), the precipitate is washed twice with 5 ml cold, 50% saturated ammonium sulphate. The precipitate is then dissolved in 1 ml Soluene-100 (Packard), and, after addition of 10 ml scintillation solution (Instafluor II, Packard), counted for radioactivity in a liquid scintillation counter. Anti-DNA is expressed in units per ml; one unit of anti-DNA is defined as the amount of anti-DNA that binds 30% of the input DNA under the above-described conditions (45). Mean DNA-binding of 135 normal human sera (50 µl tested) was 5±3% units per ml. Sera containing more than 10 units of anti-DNA per ml are considered to be positive.

(B) PEG-assay

The PEG-assay (39), is influenced by high levels of non-specific DNA-binding in normal sera. We were able to demonstrate that this is caused by binding of low density lipoproteins (LDL) to DNA (41). The non-specific DNA-binding can be eliminated by addition of excess polyanion dextran sulphate (DXS) to the incubation mixture (41).

The assay is performed as follows: 50 µl of (a dilution of) serum, 50 µl of a solution of 1.6 mg/ml HGG, 50 µl of a solution of 0.2 mg/ml DXS (Pharmacia Fine Chemicals AB, Sweden), 50 µl of a solution of 2 µg/ml (^{3}H)-PM2-DNA and 200 µl of a solution of 7% polyethylene glycol (PEG 6000; Koch-Light Labs, Colnbrook, U.K.) are mixed together and initially incubated at 37°C for 1 hr, and then at 4°C for 2 hr. All dilutions are made in PBS. After centrifugation for 15 min at 3000 g, 200 µl of the supernatant are taken, dissolved in 8 ml of NE-260$_{sp}$ (New England Nuclear Corp.) and counted for radioactivity. The mean DNA-binding of 200 normal sera (50 µl tested) was 6±2% units per ml (48). A sample is said to contain one unit of anti-DNA if it binds 30% of 100 ng DNA under the above mentioned conditions. It should be noted that, upon testing murine monoclonal antibodies to DNA, the DXS was replaced by PBS in the incubation mixture.

(C) IFT on *Crithidia luciliae*

We use this method as it was originally described (30), but with some minor modifications (31). Sera are serially diluted in PBS, starting at 1:10, and 50 µl of each dilution are incubated for 30 min, at room temperature, with a spot of *Crithidia luciliae*. After washing the slide for 30 min with PBS, 0.5 ml of fluorescein isothiocyanate (FITC) conjugated anti-immunoglobulin antiserum is layered over the slide, followed by incubation for 30 min at room temperature. The slide is washed with PBS and mounted, using a solution of 65% (w/v) sucrose in PBS, pH 8.2, in which 0.5 µg/ml propidium iodide (Calbiochem) is dissolved. Fluorescence of kinetoplasts indicates anti-dsDNA activity and the last dilution giving a positive fluorescence is used as a titre of the tested sample. The propidium iodide in the mounting solution acts as a counterstain to facilitate kinetoplast localization. A KP560 barrier filter (Leitz) is used to separate propidium iodide fluorescence from FITC fluorescence. Reading is performed at 500x with a standard fluorescence microscope, equipped with incident illumination using a Xenon arc lamp for excitation, and a 50/1.00 water immersion objective.

(D) Anti-DNA ELISA

As discussed in the previous chapter, many different reaction conditions are used in ELISA systems by various authors (44), yet all authors claim high SLE specificity of the technique.

In our experience, optimal results are obtained using the following protocol: 96-well micro-ELISA plates (Dynatech) are precoated for 2 hr with protamine sulphate (Organon, Oss, The Netherlands; 150 µl per well of a solution of 0.5 mg/ml in distilled water), washed with PBS (3 times) and with distilled water (3 times) and coated overnight at room temperature with a solution of 2 µg/ml calf thymus DNA (Worthington Biochemicals; 150 µl per well) in PBS. The plates are washed again (3 times with PBS and 3 times with distilled water; it is important to note that addition of Tween to washing buffers used after protamine sulphate precoating or DNA coating will result in partial elution of the coating from the wells); they are then incubated for 2 hr at 37°C with 100 µl of serum (or, in the case of hybridomas, culture supernatant or ascites), diluted in PBS containing 2% bovine serum albumin (BSA; Povite, Oss, The Netherlands) and 0.02% Tween-20 (Baker Chemicals). After washing the plates (3 times with PBS containing 0.02% Tween-30, and 3 times with distilled water), 100 µl of an appropriate dilution of horse-radish peroxidase conjugated antiserum is added to the wells. To test patients' sera, we use rabbit-anti-human IgG and rabbit-anti-human IgM, obtained from Dako Immunoglobulins A/S. Incubation was allowed to take place for 1 hr at 37°C, after which the plates were washed again.

Murine anti-DNA is assayed somewhat differently; after incubation of anti-DNA with the plates, plates are washed and incubated with goat-anti-mouse immunoglobulin antiserum for 1 hr at 37°C, then for 30 min at 37°C with a dilution of complexes composed of monoclonal anti-horseradish peroxidase (anti-HRP) and HRP (PAP-complexes). Dilutions are made in PBS containing 10% normal goat serum and 0.02% Tween-20.

The completion of the assay is the same whether human or murine antibodies are assayed: 100 µl of a solution of 100 µg/ml 3,5,3',5'-tetramethylbenzidine (TMB, Merck) in 0.11 M sodium acetate, pH 5.5, supplemented with 0.003% H_2O_2 are added to each well. After 15 min the reaction is stopped by addition of an equal volume of 2 M H_2SO_4 to the wells. Plates are read at 450 nm in a Titertek Multiskan. One ELISA-unit (EU) is defined as the reciprocal dilution giving an absorption of 1.0. Samples containing more than 10 EU/ml are considered positive.

Correlations between the Four Assays

Upon introduction of new techniques, it is necessary for these to be compared with existing techniques. To do so, most people use sera of patients with well-defined disease states. Comparing the Crithidia test with the Farr-assay in this way (i.e. by using sera of patients with SLE), a rather good correlation has been described (18, 24, 30, 32-38). The same holds good for comparisons of the anti-DNA ELISA with either the Crithidia-test or the Farr-assay (44, 50) and for comparisons of the PEG-assay with the Crithidia-test (43) and the Farr-assay (39, 40, 42).

In order to illustrate this, mutual comparisons of anti-DNA levels, detected by the anti-DNA ELISA, the Crithidia-test, the Farr-assay and the PEG-assay are shown in Fig. 1 - Fig. 6.

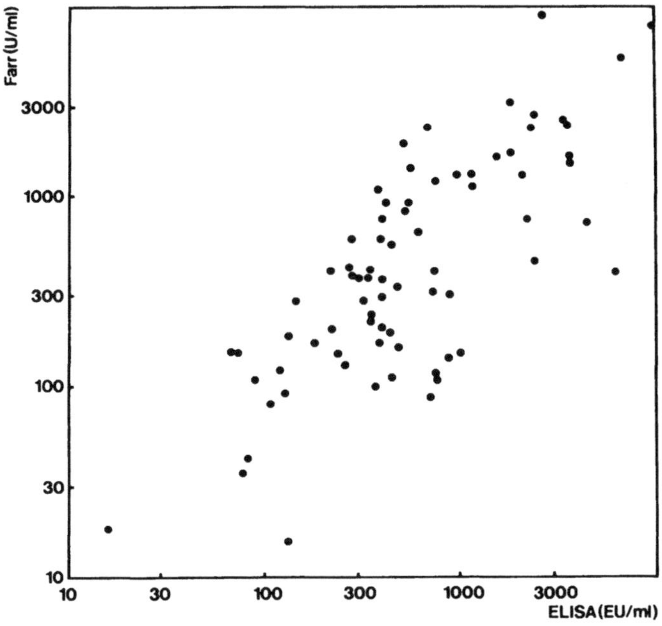

Fig. 1. Correlation between the anti-DNA ELISA and the Farr-assay

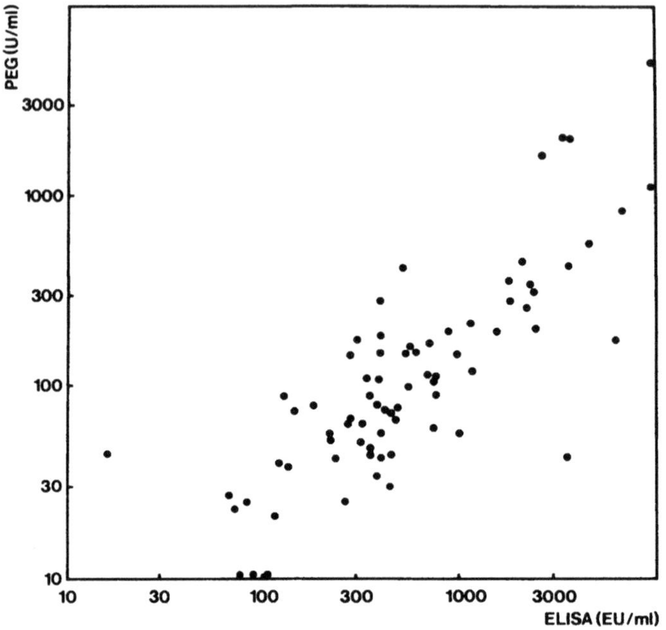

Fig. 2. Correlation between the anti-DNA ELISA and the PEG-assay

Sera of patients with defined SLE (according to the revised criteria of the American Rheumatism Association, 51) were selected in the presence of anti-DNA by means of the Farr-assay. Anti-DNA contents of these sera were measured by means of each of the previously mentioned techniques. Results obtained with one assay, on a double log scale, have been plotted against results of all other assays. Correlations between the assays were calculated by linear regression analysis and by Kendall's Rank Correlation Test, and are summarized in Table 1. From Figs. 1-6 and Table 1, it will be clear that high correlations between the various assays are found, if sera of patients with well-defined SLE are used.

Discrepancies between the Four Assays

Analysis of sera from patients with defined SLE demonstrated a good specificity of both the Farr-assay and the IFT on *Crithidia luciliae* for SLE (38). Notwithstanding the fact that the sensitivity of the Crithidia-test and the Farr-assay has been described to be of the same order of magnitude (24), we noticed a large discrepancy between the Farr-assay and the Crithidia-test upon screening of 14,417 sera sent to our institute for routine anti-DNA determination; only 37% of the IFT-positive sera were found to be Farr-positive (Table 2, taken from ref. 43). Both Taylor and co-workers (39, 40) and ourselves (41, 42) have shown that DNA/anti-DNA interactions of low avidity are dissociated in the Farr-assay, as a result of the employed ammonium sulphate precipitation. The PEG-assay lacks such dissociating reaction conditions, and indeed, we found 68% of the IFT-positive/Farr-negative sera to be PEG-positive.

Additional experiments led us to the conclusion that the deviating sera contained anti-DNA of low avidity, which is detected by the Crithidia-test and the PEG-assay, but not by the Farr-assay (41-43). Testing Crithidia-positive/PEG-positive/Farr-negative sera (i.e. sera containing predominantly anti-DNA of low avidity) in the anti-DNA ELISA, we were able to show that the ELISA also detects anti-DNA of low avidity (50). The discrepancy between the Farr-assay and the Crithidia-test is thus predominantly based on the fact that low avidity anti-DNA cannot be detected by the Farr-assay. Yet, upon development of the PEG-assay, a new discrepancy was observed, between the Crithidia-assay and the PEG-assay, as a significant number of sera were found to be PEG-positive but Crithidia-negative (42, 43).

Introduction of the ELISA to detect such antibodies was attended with some more discrepancies (50). To illustrate this problem, results obtained with the Farr-assay, the PEG-assay, the Crithidia-test and the ELISA on 289 sera referred to our institute for routine anti-DNA determination, are shown in Table 3. Given the good correlation between the various assays

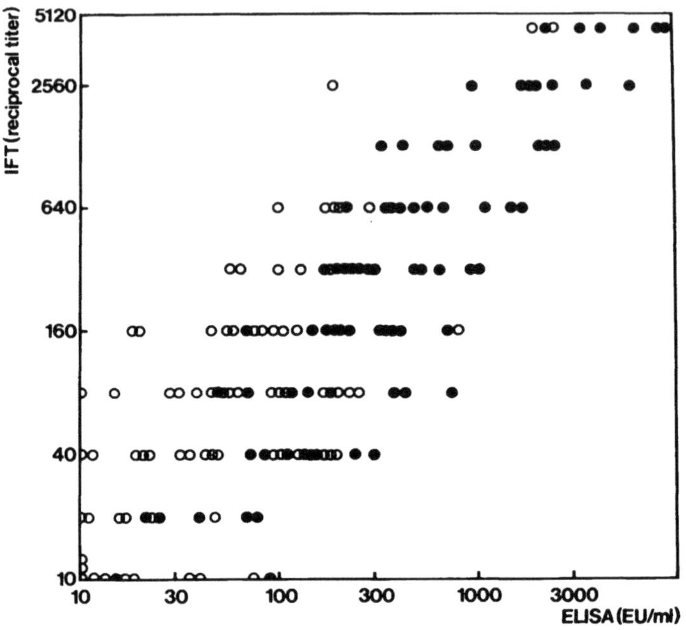

Fig. 3. Correlation between the anti-DNA ELISA and the Crithidia-test for IgG anti-DNA (●-●) and IgM anti-DNA (o-o).

Table 1. Correlations between four different anti-DNA assays. Anti-DNA levels were measured with the Farr-assay, the PEG-assay, the Crithidia-test and the anti-DNA ELISA. Correlations were calculated on a Sinclair ZX Spectrum microcomputer, using linear regression analysis and Kendall's Rank Correlation Test - (taken from 50, and Smeenk et al., unpublished results)

Anti-DNA assay	PEG-assay	Crithidia-test	Anti-DNA ELISA
Farr-assay	r=0.673(p<0.001)*	r=0.847(p<0.001)	r=0.673(p<0.001)
	τ=0.626(p<0.0001)+	τ=0.693(p<0.0001)	τ=0.530(p<0.0001)
PEG-assay	-	r=0.829(p<0.001)	r=0.622(p<0.001)
	-	τ=0.693(p<0.0001)	τ=0.631(p<0.0001)

Table 1. (continued)

Anti-DNA assay	PEG-assay	Crithidia-test	Anti-DNA ELISA
Crithidia-test	-	-	r=0.702(p<0.001) τ=0.659(p<0.0001)

* Coefficient of correlation by linear regression analysis

+ Kendall's Rank correlation coefficient

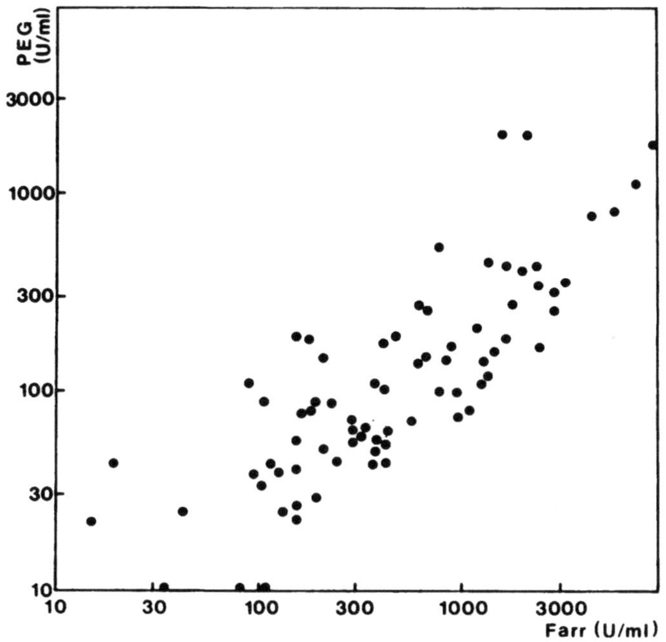

Fig. 4. Correlation between the Farr-assay and the PEG-assay.

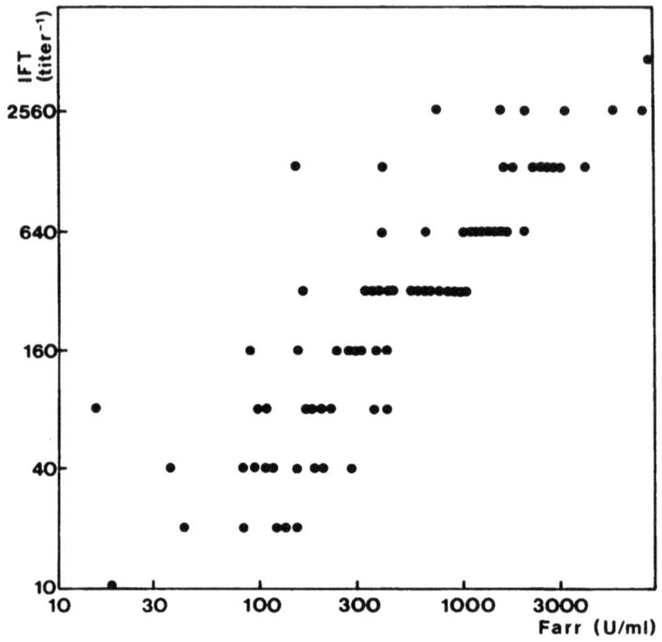

Fig. 5. Correlation between the Farr-assay and the Crithidia-test.

Table 2. Screening of 14,417 patients' sera in the presence of antibodies to dsDNA (from 43).

Anti-DNA assay	# of sera reactive pos(itive)/neg(ative)		
Crithidia-assay		1260 pos	
Farr-assay	470 pos	790 neg	
PEG-assay		535 pos	255 neg

(Table 1), one would not expect that, out of 16 possible reaction combinations, 11 actually occurred. Most of the 158 sera tested were found to be negative with all techniques, but of the remaining 131 sera, only 22 reacted positively in all four assays. Another 22 sera were Farr-negative, but positive with the other three techniques. This discrepancy is explained on the basis of avidity differences. The observation that we also found sera reacting negatively with the PEG-assay, but positively with the other three techniques may be explained by the finding that complexes need to be large to be determined with the PEG-assay (41). It is possible that anti-DNA of these patients forms small complexes with DNA, perhaps due to being of extremely high avidity. Two other rather large groups were also observed; 33 sera were found positive with the PEG-assay only and 23 sera with the ELISA only.

Overall, it looks as if the PEG-assay and the ELISA are the most sensitive techniques; anti-DNA was detected with these assays in 94 and 93 sera respectively, whereas the Crithidia-test detected anti-DNA in 59, and the Farr-assay in 30 sera.

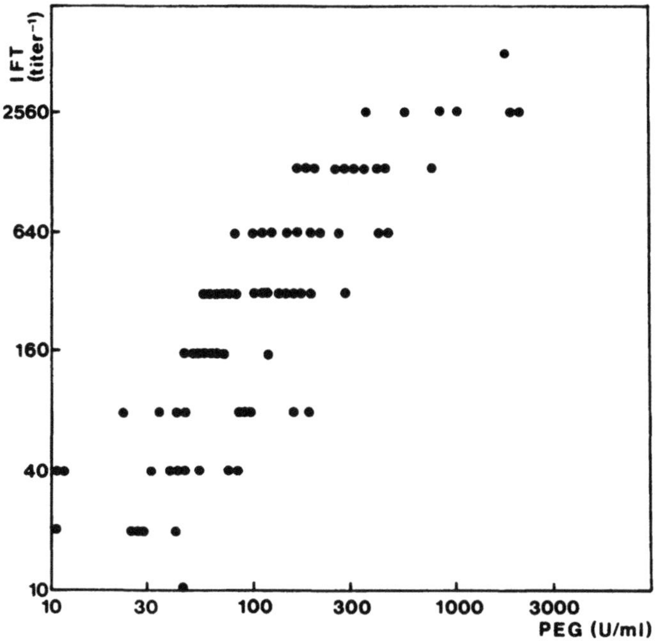

Fig. 6. Correlation between the PEG-assay and the Crithidia-test.

Table 3. Comparison of four anti-DNA assays on 289 sera referred to our institute for routine anti-DNA determination(50).

PEG-assay	ELISA	Crithidia-test	Farr-assay	Number of sera
+*	+	+	+	22
+	+	+	-	22
-	+	+	+	7
+	+	-	-	15
+	-	+	-	2
-	+	+	-	3
-	+	-	+	1
+	-	-	-	33
-	+	-	-	23
-	-	+	-	3
-	-	-	-	158
94	93	59	30	289

* + indicates a positive result; - a negative result with the technique used.

Yet, even the PEG-assay and the ELISA only detect anti-DNA in 71% of the sera found positive with any assay; hence, although they are the most sensitive techniques, they still miss anti-DNA detected by the Crithidia-test and/or the Farr-assay. This indicates that sensitivity differences between the assays can only partly explain the observed discrepancies, although assay sensitivity is thought to play a key role for sera that were found PEG- and ELISA-positive, but Farr- and Crithidia-negative. Contradictory to this is the finding of sera (n=3) that are Crithidia-positive but PEG- and ELISA-negative, sera (n=2) that are ELISA-negative but Crithidia- and PEG-positive, and sera (n=3) reacting PEG-negative, but IFT- and ELISA-positive. However, we are dealing here with very small numbers of sera.

The ELISA and the Crithidia-test are the only 2 techniques incorporating an inherent check on the Ig-nature of anti-DNA.

These assays do not detect non-specific DNA-binding, provided conjugates of high quality and good specificity are used; however, in both radioimmunoassays, non-specific DNA-binding cannot be discriminated from immunoglobulin binding. In the PEG-assay, non-specific binding was found to play an important role; it could, nevertheless, be prevented by the addition of DXS to the incubation mixture (41). Notwithstanding this, sera reacting positively in the PEG-assay only, should be checked as to whether they really contain Ig reactive with DNA, or just higher amounts than normal sera of non-Ig DNA-binding material (Clq and/or LDL), demanding more DXS to prevent binding. In the Farr-assay, we do sometimes encounter sera that react positively in the absence, but negatively in the presence of DXS, whereas DXS in general has no effect on DNA-binding in the Farr-assay (41). DNA-binding of such sera sediments at 11 S upon sucrose-gradient analysis, and, therefore, is assumed to be due to Clq, although we cannot explain why this is never found with control sera (Smeenk et al., unpublished results).

Because patients' sera will always contain a spectrum of many distinct antibodies with varying specificities and avidities, they are not in fact very suited to aid unravelling the basis of the varying behaviour of antibodies to DNA in the different assays. One would expect more clear results working with monoclonal antibodies. As has been performed by others, we too have screened the clones obtained from fusion of spleen cells from SLE mice with SP2/0 myeloma cells with the anti-DNA ELISA on anti-DNA production (52). From several fusions we obtained a total of 40 different anti-DNA producing hybridomas, of which not one was found to be Farr-positive. Reaction patterns of these monoclonals in the various anti-DNA assays are shown in Table 4 (from ref. 52). Clearly, all monoclonals are of low avidity. There is a rather good correlation between the Crithidia-test and the PEG-assay; 81% of the Crithidia-positive clones were also PEG-positive. Yet, from the Crithidia-negative clones, 38% were found to be PEG-positive. From these results, one would hypothesize that the ELISA has a much greater sensitivity than any of the other assays, or detects antibodies of such a low avidity that they are missed by all other assays.

Table 4. Reaction patterns of 40 anti-DNA hybridomas in different anti-DNA-assays (52).

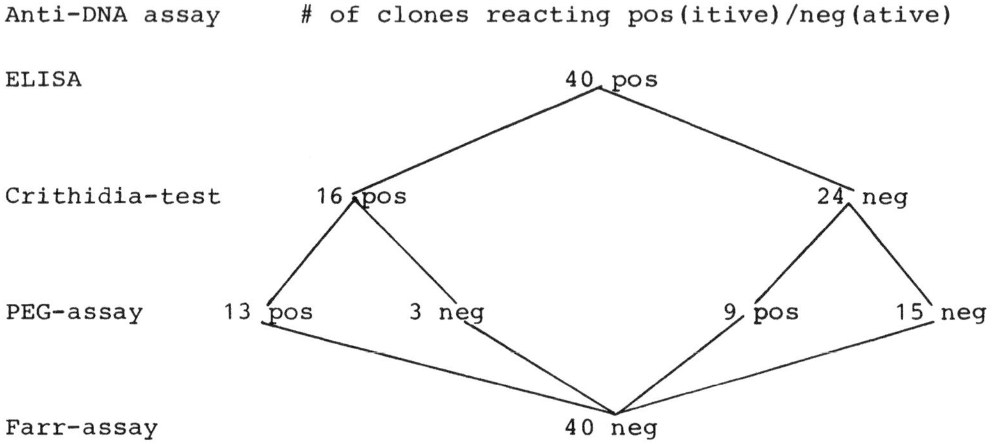

Concluding Remarks

In this chapter, we have compared four different anti-DNA assays, two of which are radioimmunoassays. Both from the literature and from our own results, we are inclined to say that the Farr-assay has the greatest SLE-specificity, provided purely dsDNA is used as antigen. Hence, for purely diagnostic use, this should be the preferred technique.

However, low avidity anti-DNA is not detected by this method. From our results, anti-DNA of low avidity seems to be less specific for SLE than anti-DNA of high avidity. Both with the PEG-assay as well as with the ELISA, we sometimes detect low levels of anti-DNA in related disorders such as rheumatoid arthritis, thyroid autoimmune disease, Sjögren's syndrome, scleroderma, autoimmune gastritis, autoimmune liver disease and myasthenia gravis (48, 50). On the other hand, simultaneous detection of antibodies of high avidity and low avidity might allow discrimination between subsets of SLE patients.

Calculating a ratio between Farr- and PEG-results, and relating obtained figures with disease symptoms displayed by the patients studied, a significantly higher incidence of lupus nephritis was found (53) in patients with high avidity antibodies and an increased incidence of central nervous system involvement in patients with a relative preponderance of low avidity anti-DNA. In a prospective study of 441 non-SLE patients in which anti-DNA could be detected by the Farr-assay, it was found (54) that, after 5 years, 85% of these patients had developed SLE. Patients with relatively high avidity anti-DNA were more prone to develop SLE than patients with relatively low avidity anti-DNA. It is difficult, therefore, to draw conclusions using only one assay in SLE; it would probably be best to use several assays at the same time.

A remarkable finding was that Crithidia-positive/Farr-negative sera, when tested in the Farr-assay with singlestranded DNA (ssDNA) as antigen, often reacted positively (49). Most probably, anti-DNA that interacts with low avidity with dsDNA, will give an interaction of much higher avidity with ssDNA, because of the much greater flexibility of ssDNA strands compared with dsDNA. It has been shown how important flexibility is in terms of allowing multipoint attachment - and thus high avidity binding - of antibodies to DNA (55). Using an anti-DNA assay based on the inhibition of ethidium-bromide intercalation with dsDNA, to which anti-DNA is bound, it was found (56) that anti-ssDNA also inhibits this intercalation. It was concluded that so-called anti-ssDNA may in fact be directed to antigenic determinants that are also available for binding in the dsDNA molecule. It has been shown (57) that reactivity of antibodies to dsDNA with ssDNA is mainly due to reactivity with internal duplex structures in the ssDNA.

From all these results, we conclude that most of what is called anti-ssDNA will probably with low avidity react with dsDNA. However, not only will immunoglobulin react with ssDNA with higher avidity than with dsDNA, but also this will be the case for other serum proteins. Therefore, assays using ssDNA will suffer from all sorts of non-specific DNA binding by serum proteins. Assays such as the ELISA and Crithidia assay overcome this problem by incorporating an intrinsic check on the Ig-character of DNA-binding material. In assays without such a check, the use of purely dsDNA must be combined with either a mechanism selecting for high avidity DNA-binding (such as the Farr-assay) or a means to prevent non-specific binding to DNA (as accomplished by addition of DXS to the PEG-assay). With all the above mentioned restrictions in mind, it must be possible to measure antibodies to DNA in such a way that they are specific for SLE, and available for use in following individual patients during the course of their disease.

Acknowledgements

The author wishes to thank Dr. T. Swaak for sera and clinical information on the patients, Dr. L. Aarden for valuable discussions and Drs. Aarden and Van der Schaal for critically reading the manuscript. Without the expert skill of Yvonne Kluyt and Annette Lochtenberg, this typescript would never have existed.

References

1. Hughes, G.R.V.: Systemic lupus erythematosus. In: "Connective Tissue Diseases", Blackwell Scientific Publications, Oxford, pp. 3-65 (1977).
2. Oster, W.: On the visceral complications of erythema exudativum multiforme. Am. J. med. Sci. 110, 629-639 (1895).
3. Stollar, D.B.: Nucleic acid antigens. In: "The Antigens", vol. I, ed. Sela, M., Academic Press, New York, pp. 1-85 (1973).
4. Krishnan, C., Kaplan, M.H.: Immunopathologic studies of SLE. II. Antinuclear reaction of gammaglobulin eluted from homogenates and isolated glomeruli of kidneys from patients with lupus nephritis. J. clin. Invest. 46, 569-579 (1967).

5. Koffler, D., Agnello, V., Thoburn, R., Kunkel, H.G.: SLE, prototype of immunecomplex nephritis in man. J. exp. Med. 134, 169S-179S (1971).

6. Winfield, J.B., Koffler, D., Kunkel, H.G.: Role of DNA/anti-DNA complexes in the immunopathogenesis of tissue injury in SLE. Scand. J. Rheumatol. suppl. 11, 59-62 (1975).

7. Bruneau, Ch., Benveniste, J.: Circulating DNA/anti-DNA complexes in SLE: Detection and characterization by ultracentrifugation. J. clin. Invest. 64, 191-198 (1979).

8. Lafer, M., Rauch, J., Andrzejewski, C. Jr., Mudd, D., Furie, B., Schwartz, R.S., Stollar, B.D.: Polyspecific monoclonal autoantibodies reactive with both polynucleotides and phospholipids. J. exp. Med. 153, 897-909 (1981).

9. Shoenfeld, Y., Rauch, J., Massicotte, H., Datta, S.K., André-Schwartz, J., Stollar, B.D., Schwartz, R.S.: Polyspecificity of monoclonal lupus autoantibodies produced by human-human hybridomas. New Engl. J. Med. 308, 414-420 (1983).

10. Jacob, L., Tron, F., Bach, J.F., Louvard, D.: A monoclonal anti-DNA antibody also binds to cell-surface protein(s). PNAS USA 81, 3843-3845 (1984).

11. Faaber, P., Capel, P.J.A., Rijke, G.P.M., Vierwinden, G., Van de Putte, L.B.A., Koene, R.A.P.: Crossreactivity of anti-DNA antibodies with proteoglycans. Clin. exp. Immunol. 55, 502-508 (1984).

12. Eilat, D.: Crossreactions of anti-DNA antibodies and the central dogma of lupus nephritis. Immunol. Today 6, 123-127 (1985).

13. Seligmann, M.: Serologie dans le sérum de malades atteints de lupus erythematosus disseminé. CR Acad. Sci. (Paris) 245, 243-255 (1957).

14. Ceppelini, R., Polli, E., Celada, F.: A DNA-reacting factor in serum of a patient with lupus erythematosus diffusus. Proc. Soc. exp. Biol. Med. 96, 572-574 (1957).

15. Miescher, P., Strassle, R.: New serologic methods for the detection of the L.E. factor. Vox Sang. 2, 283-295 (1957).

16. Robbins, W.C., Holman, H.R., Deicher, H.R., Kunkel, H.G.: Complement fixation with cell nuclei and DNA in lupus erythematosus. Proc. Soc. exp. Biol. Med. 96, 575-579 (1957).

17. Feltkamp, T.E.W.: Editor: "The Significance of the Determination of anti-DNA and DNA/anti-DNA complexes". Scand. J. Rheumatol. suppl. 11 (1975).

18. Maini, R.N., Holborrow, E.J.: Editors: "Detection and Measurement of anti-DNA Antibodies". Ann. Rheum. Dis. 36, suppl. 1 (1977).

19. Stollar, B.D.: The specificity and applications of antibodies to helical nucleic acids. CRC. Critical Reviews in Biochemistry 3, 45-69 (1975).

20. Hasselbacher, P., Leroy, E.C.: Serum DNA-binding activity in healthy subjects in rheumatic disease. Arthr. Rheum. 17, 63-71 (1974).

21. Jain, S., Markham, R., Thomas, H.C., Sherlock, S.: Double-stranded DNA-binding capacity of serum in acute and chronic liver disease. Clin. exp. Immunol. 26, 35-41 (1976).

22. Grennan, D.M., Sloan, D., Behan, A., Carson, W.D.: Clinical significance of antibodies to native DNA as measured by a DNA-binding technique. Ann. Rheum. Dis. 36, 30-33 (1977).

23. Permin, H., Halberg, P., Christiansen, P.: Antibodies against dsDNA in patients with connective tissue diseases. Acta med. Scand. 203, 61-65 (1978).

24. Aarden, L.A., Lakmaker, F., De Groot, E.R., Swaak, A.J.G., Feltkamp, T.E.W.: Detection of antibodies to DNA by radioimmunoassay and immunofluorescence. Scand. J. Rheumatol. suppl. 11, 12-19 (1975).

25. Holian, J., Griffiths, I.D., Glass, D.N., Maini, R.N., Scott, J.T.: Human anti-DNA antibody: reference standards for diagnosis and management of SLE. Ann. Rheum. Dis. 34, 438-443 (1975).

26. Medof, M.E., Locker, J.D., Bennett, R.M.: Antibodies to DNA: false-positive results. Arthr. Rheum. 19, 963-964 (1976).

27. Wold, R.T., Young, F.E., Tan, E.M., Farr, R.S.: Deoxyribonucleic acid antibody, a method to detect its primary interaction with deoxyribonucleic acid. Science 161, 806-810 (1968).

28. Aarden, L.A., Lakmaker, F., Feltkamp, T.E.W.: Immunology of DNA. I. The influence of reaction conditions on the Farr-assay as used for the detection of anti-dsDNA. J. Immunol. Meth. 10, 27-37 (1976).

29. Aarden, L.A., Lakmaker, F., Feltkamp, T.E.W.: Immunology of DNA. II. The effect of size and structure of the antigen on the Farr-assay. J. Immunol. Meth. 10, 39-48 (1976).

30. Aarden, L.A., De Groot, E.R., Feltkamp, T.E.W.: Immunology of DNA. III. *Crithidia luciliae*: a simple substrate for the detection of anti-dsDNA with the immunofluorescence technique. Ann. N.Y. Acad. Sci. 254, 505-509 (1975).

31. Aarden, L., Smeenk, R.: Measurements of antibodies specific for DNA. In: "Immunological Methods", vol. II, Eds. Lefkovitz, I., Pernis, B., Academic Press, New York, pp. 75-82 (1981).

32. Stingl, G., Meingassner, J.G., Swelty, P., Knapp, W.: An immunofluorescence procedure for the determination of antibodies to native dsDNA and of circulating DNA/anti-DNA complexes. Clin. Immunol. Immunopathol. 6, 131-137 (1976).

33. Slater, N.G.P., Cameron, J.S., Lessof, M.M.: The *Crithidia luciliae* kinetoplast fluorescence test in SLE. Clin. exp. Immunol. 25, 480-488 (1976).

34. Crowe, W., Kushner, I.: An immunofluorescence method using *Crithidia luciliae* to detect antibodies to dsDNA. Arthr. Rheum. 20, 811-817 (1977).

35. Sontheimer, R.D., Gilliam, J.N.: An immunofluorescence assay for dsDNA antibodies using the *Crithidia luciliae* kinetoplast as a dsDNA substrate. J. Lab. clin. Med. 91, 550-556 (1978).

36. Monier, J.C., Perraud, M., Ricard, F., Gioud, M.: Comparison of three techniques for the detection of antibodies to dsDNA: immunofluorescence on *trypanosoma gambiense*, immunofluorescence on *Crithidia luciliae* and radioimmunoassay using the Farr-technique. Ann. Immunol. (Inst. Pasteur) 129C, 463-469 (1978).

37. Ballou, S.P., Kushner, I.: Immunochemical characteristics of antibodies to DNA in patients with active SLE. Clin. exp. Immunol. 37, 58-64 (1979).

38. Swaak, A.J.G., Groenwold, J., Aarden, L.A., Feltkamp, T.E.W.: Detection of anti-dsDNA as a diagnostic tool. Ann. Rheum. Dis. 40, 45-49 (1981).

39. Riley, R.L., McGrath, H. Jr., Taylor, R.P.: Detection of low-avidity anti-DNA antibodies in SLE. Arthr. Rheum. 22, 219-225 (1979).

40. Riley, R.L., Addis, D.J., Taylor, R.P.: Stability of DNA/anti-DNA complexes. II. Salt-lability and avidity. J. Immunol. 124, 1-5 (1980).

41. Smeenk, R., Aarden, L.: The use of PEG-precipitation to detect low-avidity anti-DNA antibodies in SLE. J. Immunol. Meth. 39, 165-180 (1980).

42. Smeenk, R., Van der Lelij, G., Aarden, L.: Avidity of antibodies to dsDNA. Comparison of IFT on *Crithidia luciliae*, Farr-assay and PEG-assay. J. Immunol. 128, 73-78 (1982).

43. Smeenk, R., Van der Lelij, G., Aarden, L.: Measurement of low-avidity anti-dsDNA by the *Crithidia luciliae* test and the PEG-assay. Clin. exp. Immunol. 50, 603-610 (1982).

44. Smeenk, R.: Detection of antibodies to DNA by enzyme-linked immunosorbent assay (ELISA). This issue, pp. 121-143 (1986).

45. Aarden, L.A., Lakmaker, F., De Groot, E.R.: Immunology of DNA. IV. Quantitative aspects of the Farr-assay. J. Immunol. Meth. 11, 153-163 (1976).

46. Smeenk, R., Duin, T., Aarden, L.: Influence of pH on the detection of low- and high-avidity anti-dsDNA. J. Immunol. Meth. 55, 361-373 (1982).

47. Rubin, R.L., Lafferty, J., Carr, R.I.: Re-evaluation of the ammonium-sulphate assay for DNA antibody. Arthr. Rheum. 21, 950-954 (1978).

48. Smeenk, R., Van der Lelij, G., Swaak, T., Groenwold, J., Aarden, L.: Specificity to SLE of antibodies to dsDNA measured with the PEG precipitation assay. Arthr. Rheum. 25, 631-638 (1982).

49. Aarden, L., Smeenk, R.: Immunochemical properties of antibodies to DNA and their influence on detection methods. In: "Antibodies to Nuclear Antigens", Eds. Kalden, J.R., Feltkamp, T.E.W., Excerpta Medica, Amsterdam, pp. 23-28 (1982).

50. Smeenk, R.: Low-avidity antibodies to dsDNA. Thesis, Rodopi, Amsterdam (1982).

51. Tan, E.M., Cohen, A.S., Fries, J.F.: The 1982 revised criteria for the classification of SLE. Arthr. Rheum. 25, 1271-1275 (1982).

52. Smeenk, R., Westgeest, T., Duin, T., Van der Brink, H.: Crossreactivity of monoclonal antibodies to DNA: related to avidity ? Protides Biological Fluids vol. XXXIII, Ed. Peeters, H. (in press, 1985).

53. Swaak, T., Smeenk, R.: Clinical differences between SLE patients in relation to the avidity of their anti-dsDNA. Protides Biological Fluids vol. XXXIII, Ed. Peeters, H. (in press, 1985).

54. Swaak, T., Smeenk, R.: Detection of anti-dsDNA as a diagnostic tool: a prospective study in 441 non-SLE patients with anti-dsDNA antibodies. Ann. Rheum. Dis. 44, 245-251 (1985).

55. Aarden, L.A., Lakmaker, F., De Groot, E.R.: Immunology of DNA. VI. The effect of mercaptans on IgG and IgM anti-dsDNA. J. Immunol. Meth. 16, 143-153 (1977).

56. Shephard, J.D., Fritzler, M.J., Watson, J.I., Van de Sande, J.H.: The binding of anti-DNA antibodies as measured fluorometrically by ethidium-bromide. J. Rheumatol. 5, 391-396 (1978).

57. Stollar, B.D., Papalian, M.: Secondary structure in denatured DNA is responsible for its reaction with anti-native DNA antibodies of SLE sera. J. clin. Invest. 66, 210-215 (1980).

LIPOSOME IMMUNOASSAY

D. Monroe
Department of Medicine, Infectious Diseases and Connective Tissues Sections, University of Tennessee, Center for the Health Sciences, Memphis, Tennessee 38163, U.S.A.

Introduction

Numerous immunoassays have recently been developed for the rapid determination of specific substances in biological fluids. Among these is the highly sensitive, specific, fast, and easily performed liposome immunoassay (LIA). This assay utilizes artificial membrane lipid vesicles called liposomes which are engineered to bind certain molecules and release encapsulated reagents for their rapid detection.

Under carefully controlled conditions, liposomes readily form when phospholipids are mixed in an aqueous solution. Such microcapsular micelles are composed of internal aqueous compartments surrounded by closed concentric lipid bilayers. Drugs, dyes, proteins, enzymes, various salts, or other desired molecules can be conveniently carried in these aqueous spaces. Receptor molecules, antigens, and antibodies are often attached to the outer liposomal surface while other chemicals may be membrane incorporated as part of the lipid bilayer.

Because of these properties, liposomes have many applications which include the use of several new therapeutic and diagnostic products. These vesicles prove excellent vehicles for delivering drugs, hormones, enzymes, genetic material, and fluorescent markers to specific cellular targets. Liposome carriers can be used in treating cancer, diabetes, arthritis, phlebitis, and a variety of infectious diseases. Natural targeting of liposomes to macrophages greatly increases the adjuvant

Immunoassay Technology, Vol. 2

activity of encapsulated muramyl dipeptides that stimulates immune responses (1).

Attaching macromolecules to liposomes provides polyvalent receptor targets that bind specific particles with greater avidity for better resolution in agglutination assays. Diagnostic tests for syphilis, hepatitis B, or mononucleosis will be greatly improved by using several liposome enhanced agglutination protocols. Liposome-based immunoassays for measuring thyroid hormone T_4 in screening congenital hypothyroidism disorders (2) and early detection of systemic lupus erythematosus are providing more accurate, cheaper, and faster methods for making a diagnosis (3, 4).

Composed of materials similar to cell membranes, liposomal phospholipid dispersions serve well as model systems suitable for natural membrane function investigations. The arrangement of phospholipid molecules, however, can be altered in the presence of cytotoxins or fixed complement resulting in content loss. Materials released can be used to perform specific functions and for quantitating the immunological reactions involved in liposomal lysis. Liposomes are easily engineered for binding specific target molecules and release certain carrier substances when binding takes place with subsequent rupture. They offer precise and unique LIA systems which compare well with other established immunoassays such as enzyme-linked immunosorbent assay (Table 1) or ELISA (5). Advanced liposome technology (6), automation, and industrial scale up production now possible (7) will enable LIAs to become important immunological research tools as well as essential parts of valuable diagnostic test kits.

Vesicle Formation

Liposomes can readily be made from a mixture of chemicals that includes lipids and phospholipids. These fatty molecules dissolved in organic solvents are deposited as thin films on boro-

Table 1. Comparison of ELISA and LIA

	ELISA	LIA
Sensitivity	High (10^{-14}mol)	High (10^{-15}mol)
Specificity	High (Antigens/ Haptens/Antibodies)	High (Antigens/ Haptens/Antibodies)
Speed	16 - 20 hours	1 - 2 hours
Reaction	Heterogeneous	Homogeneous/Heterogeneous
Automation	Excellent	Ideal
Reagents	Long shelf life	Very stable
Assay Temperature	Controlled	Ambient

silicate glass round bottom flask or test tubes by means of rotary evaporation. When water is added for rehydration, the lipids spontaneously form bilayer membranes consisting of an outer aqueous soluble hydrophilic portion and a hydrophobic water insoluble interior. If the test tube or flask is gently shaken, the membrane is temporarily disrupted and several closed multilayered lipid vesicles or micelles form rapidly (3). Each resulting liposome will then consist of a single or many onionskin-like multilamellar lipid layers separating individual aqueous compartments (Fig. 1).

Vesicle characteristics are largely determined by many influencing factors present during their formation. These may include: length of time required for complete hydration, methods used for phospholipid and lipid resuspension, the amount of dried lipid film or thickness on a glass surface, temperature, pH, degree of agitation, and the different chemicals as well as their proportions used in making liposomes (1).

Different liposome sizes (25 nanometers to 25 micrometers) are obtained by varying the length of hydration time and the amount of agitation used. Aqueous phase encapsulation can be increased by as much as 50% per lipid mole if the hydration time is varied from 2 to 20 hours with gentle shaking. Smaller sized vesicles with very little aqueous space will form by decreasing the hydration time to 30 minutes with vigorous

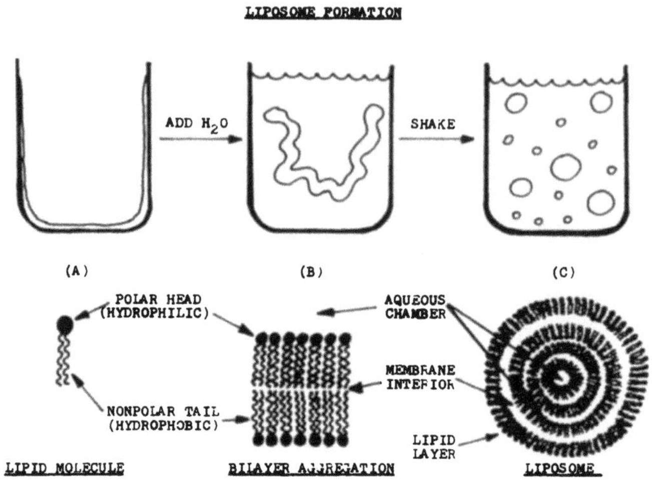

(A) LIPIDS CONSISTING OF A HYDROPHILIC POLAR HEAD (CARBOXYLATE IONS) AND NONPOLAR HYDROPHOBIC TAILS (HYDROCARBON CHAINS) ARE DRIED AS THIN FILMS ON A GLASS SURFACE. (B) WHEN WATER IS ADDED, A BILAYER AGGREGATION (MEMBRANE) FORMS WITH POLAR HEAD SURFACES EXPOSED TO THE AQUEOUS ENVIRONMENT AND NONPOLAR TAILS MAKING UP THE INTERIOR.
(C) CLOSED MULTILAMELLAR VESICLES (LIPOSOMES) CONSISTING OF MANY LIPID LAYERS SEPARATING AQUEOUS CHAMBERS FORM WITH GENTLE SHAKING.

Fig. 1

agitation. Less time is required for efficient hydration when very thin lipid films are formed on glass surfaces by rotary evaporation and gentle heating.

Aqueous phase entrapment becomes less when phosphatidyl choline (lecithin) is used in making liposomes. Such neutral lipids cause very small interlamellar spacing to take place. Charged lipids like dicetyl phosphate repels bilayers from each other and consequently increases aqueous phase entrapment. Liposome membrane integrity is preserved somewhat when lipids are dissolved in cholesterol. They virtually remain unchanged for a four month period when properly stored and were shown to be stable over 8 months when stored at $4^{o}C$ under nitrogen (5). If cholesterol is omitted as a vesicle constituent, however, liposome leakage of entrapped reagents occurs at an approximate rate of 7% per month (8). Cholesterol seems to decrease liposome permeability and small molecular vesicle leakage is reduced. Other factors influencing liposome stability include lipid composition, trapped marker, protein amount on the liposome, and coupling method employed for binding reagents to liposome surfaces (5). Anionic solute leakage is facilitated by adding lysozyme to liposome suspensions while ether or chloroform additions will increase cation permeability of the bilayers.

Large unilamellar (LUV) or small single walled vesicles (SUV),approximately 150 nanometers in diameter, are the ideal size for liposomes used in most LIAs. A variety of liposome types that vary in size and production methods are summarized in Table 2.

Table 2. Liposome types

Vesicles	Aqueous spaces	Lipid layers	Diameter (NM)	Production method	Ref.
Multilamellar (MLV)	Several	Various	500-10,000	Lipids dissolved in organic solvents and rotary evaporated. Water added with agitation.	9
Large Unilamellar (LUV)	One	One	200-1,000	Ether infusion	10, 11
				Calcium fusion	12
				Reverse-phase evaporation.	13
Small Unilamellar (SUV)	One	One	25-100	Sonication	14, 15
				Ethanol injection	16
				Detergent dialysis	17
				Reverse micelle method.	18

Single wall lipid vesicles (SUV) can be formed by mixing egg lecithin (2.4×10^{-5}mol) dissolved in 176 µl of 2:1 (v/v) chloroform-methanol with cholesterol (1.2×10^{-5}mol) dissolved in 44 µl of chloroform (8). Multilamellar liposomes (MLV) are readily produced by mixing egg lecithin, cholesterol, dicetyl phosphate, and even an appropriate antigen if desired in the molar ratios of 2:1.5:0.2:0.01 (19, 20). Gently bubbling either argon or nitrogen gas promotes chemical mixing during liposome formation and helps reduce oxidation (4). The phospholipid-cholesterol mixture when dried to a thin film can easily be rehydrated and solubilized by adding a detergent.

A 1% sodium deoxycholate solution is very efficient in solubilizing most membranes. The detergent can then be removed from solution by gel filtration or dialysis (21). However, a minimum volume (4 ml) of 200 mM non-ionic, non-denaturing octyl glucoside detergent is often added instead, followed by vigorous mixing at ambient temperature (8, 22). A reagent or marker of 5-10 mg, to be liposome encapsulated (enzymes, drugs, hormones, ions, etc.), is dissolved in the lipid-detergent mixture. When this is dialyzed against several changes of 0.05M Tris-HCl, pH 7.8 at 25°C for 24-48 hr, the detergent is removed and liposomes containing markers are reformed. The liposome dialyzed solution is then chromatographed to remove free reagents, detergent, and to fractionate the desired sized vesicles (Fig. 2).

Liposomes of approximately the same size release their contents at a constant rate resulting in more accurate LIA measurements being obtained as lysis takes place. Vesicle preparations are satisfactorily fractionated on either cross-linked agarose or dextran gel media. Equal sized homogeneous liposomes elute in the void volume of either a Sepharose, Sephadex, or Sephacryl column. The uniform vesicle suspension can be stored for several months at 4°C in 0.05 M Tris-HCl, pH 7.8 buffer with little leakage of the encapsulated reagents.

Vesicle Lysis

A suspension of immunologically sensitive liposomes will precipitate when certain antisera bind to those antigens solubilized in the vesicle bilayer membrane. Liposomal membrane sensitization

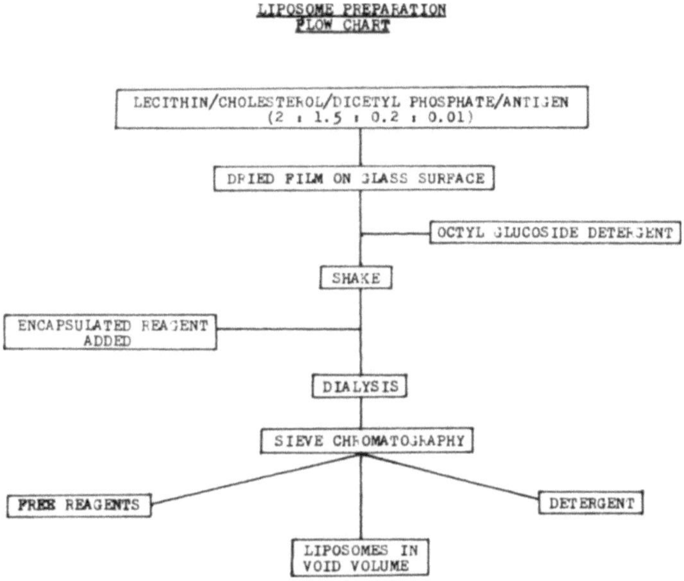

Fig. 2

is due to most major histocompatible proteins naturally exposing their antigenic determinants externally when being incorporated into the vesicle. Surface antigens can then easily become targets for type specific antibodies in immune assays (23). Appropriate antibodies quickly bind to these antigens, forming immune-complexes which co-precipitate when subsequently exposed to secondary antibodies directed against those already antigen bound. If fluorescent labelled lipid vesicles are used in the immunoassay, resulting liposomal precipitation can be conveniently quantitated by fluorescent microscopy without vesicle content loss.

Vesicle lysis can be complement mediated or simply achieved by addition of potent cytolytic agents. If complement is added together with antibodies that bind liposomal antigens, immune-complex complement fixation takes place without precipitation.

Instead, bilayer membrane rupture occurs at the complement bound sites resulting in sudden release of entrapped vesicle contents or markers. Measuring the amount of vesicle marker release can be used to determine antigen, antibody, or complement levels in immunoassay systems.

While complement dependent LIAs are simple to perform, a few disadvantages such as cost are probably responsible for this technique not being as widely accepted as it should be (22). Complement-mediated LIAs are somewhat expensive for several reasons. Large amounts of complement are often necessary for vesicle lysis and freshly prepared liposomes containing specific antigens are usually required for each separate application. Test sensitivity is often variable and limited, depending on liposome membrane integrity, antigens incorporated, antibody affinity, and complement levels present. Guinea-pig serum, often used as a complement source, is suspected to contain various kinds of non-specific liposome lysis factors. Consequently, optimal concentrations or a precise guinea-pig serum dilution must be predetermined before each test so as to avoid such factors (5). Considerably lower cost and elimination of non-specific lysis, however, is possible by performing LIAs which are independent of complement. Using cytolytic agents helps to make LIAs reasonably inexpensive as well as offering several other advantages, as summarized in Table 3.

Table 3. Liposome lysis

Complement dependent	Cytotoxin induced
Expensive due to high complement concentrations required for increasing kinetic rate.	Lower test cost due to lack of complement use and little antigen or hapten required.
Separate liposome preparations required for each LIA since specific antigens are membrane incorporated.	Surface antigens not required. Single, stable liposome preparations can be used for measuring several different analytes.
Antigen incorporation not guaranteed since lipid chemical characteristics may vary during conjugation and cause loss of membrane integrity.	Liposome membrane integrity remains stable. Antigens are conjugated only to cytotoxins. Biological activity of conjugate must be retained.

Table 3 (contd)

Complement dependent	Cytotoxin induced
Sufficient complement and type specific antibody required for liposomal surface antigen binding.	Only cytolytic agent covalently linked to free antigen or hapten required for membrane lysis.
Sensitized liposomal surface depends on number of antigen bound molecules. Test sensitivity dependent on these antigens, complement levels, and antibody affinity.	Test sensitivity limited only by antibody affinity and cytolytic potency. LIAs are very sensitive due to antigen incorporated vesicles not required for test. Use of more potent cytotoxins will increase sensitivity.
Limited test sensitivity for very low concentrations of analytes. They are often undetected due to antibody binding with overwhelming number of competing antigen molecules present on vesicle sensitized surfaces.	Lower LIA sensitivity range is possible since antigen vesicle membrane not required. Any agent causing lipid bilayer leakage can be used to release contents.
Assay sensitivity also restricted to a competitive test since it must be performed only in a homogeneous manner without a separation step.	LIA is performed in either a homogeneous competitive system or coupled to a noncompetitive heterogeneous assay requiring a separation step for increased sensitivity and removal of interference.
Several molecules are required in the complement cascade for lysis. Slow kinetics result with a 20-30 minute reaction time. Marker release is slow.	A single molecule causes lysis which is instantaneous with a 5-10 minute reaction time. Very rapid marker release with no lag time.

Phospholipases, apolipoproteins, and lipid bilayer channel formers cause vesicle lysis without the need of complement (8). When these cytolytic agents are used to cause liposome lysis, there is no need to sensitize vesicles by incorporation or membrane attachment of appropriate antigens. Instead, liposomes are prepared without antigen vesicle incorporation. This results in a single stable liposome lot that can be used in a variety of LIAs for measuring different analytes. The major requirement for these assays, however, is that biological activity must be maintained after conjugation of the cytolytic agent with an antigen or antibody. This assay approach is highly sensitive, simple, fast, versatile, and offers a means of measuring multiple analytes economically (8, 22). Using more potent cytolysins and better markers will enable the ultimate sensitivity limit of LIAs to be within the picomolar range (8).

Vesicles with 5% cardiolipin membrane incorporated will lyse within just a few minutes when added to a mixture of anti-cardiolipin antisera and complement (24). However, liposomes containing 40% beef-heart cardiolipin as part of their bilayer membrane component will rupture in the presence of divalent cations independent of complement. Either calcium or magnesium ions can cause bilayer lamellar-to-hexagonal rearrangement of liposomal membrane cardiolipin with subsequent release of encapsulated markers. Systemic lupus erythematosus (SLE) autoantibodies, however, stabilize the membrane surface by specifically binding to the cardiolipin molecules. Liposomal structural changes are then blocked, preventing membrane leakage. Such a novel liposome composition has resulted in development of a rapid LIA colorimetric test for SLE (4) which appears to be more sensitive and advanced than other SLE assays currently available.

Markers

A variety of liposomal encapsulated reagents known as labels or markers have been used by many investigators in performing different LIAs. These markers have included: dyes, enzymes, sugar molecules, ions, and electron spin labels. The type, size, or molecular weight of a marker are important factors of liposome stability (5). Fluorescein-labelled lipid vesicles can conveniently be measured by means of spectrophotometry or fluorescence microscopy (23). Immuno-precipitation of these liposomes concentrates the fluorescein dye and eliminates leakage problems

Table 4. Liposome entrapped labels

Label	Specific marker	Detection method[a]	Ref.
Dye	Fluorescein	Spectrophotometer or microscope	23, 5
Dye	Arsenazo III	Spectrophotometer	4
Enzyme	Alkaline phosphatase	Spectrophotometer	8,22
Enzyme	Horseradish peroxidase	Selective electrode	20
Sugar	Glucose	Spectrometer	25,26
Ion	Tempocholine chloride	Electron spin resonance spectrometer	2,24
Spin	5-, 10-, or 12-Doxylstearic acid	Electron spin resonance spectrometer	27

[a]Bioselective electrodes as described by Monroe (28, 29) offer alternative methods for detecting dyes, enzymes, sugar and ions.

often occurring when lysis techniques are employed. Carboxyfluorescein, a highly water soluble derivative of fluorescein, is commonly used. This self-quenching compound shows little fluorescence while encapsulated, whereas the dye becomes highly fluorescent when diluted in a large volume as a result of immune lysis (5).

Other liposomal incorporated dyes change colour under certain conditions when vesicle lysis takes place. Intact liposomes containing the red dye Arsenazo III have been used in LIAs (4) and are easily measured spectrophotometrically at a wavelength of 550 nm. When lysed in the presence of added magnesium, this liposomal released dye quickly forms a stable blue cation complex detected at 610 nm wavelength. Such a unique colour change and different absorption spectra forms the basis for quantification of LIA results.

Maximum LIA sensitivity, however, is achieved when enzymes are encapsulated as liposome markers (8). Upon vesicle lysis, enzymes are suddenly released to rapidly change thousands of substrate molecules into products. As substrate conversion continues, a detectable signal, such as a chromogenic colour change or an electronic response, results. Signal amplification depends largely on the high enzyme turnover rate, vesicle size, and the number of enzyme markers available per vesicle. Complete enzyme release is accomplished when unilamellar vesicles are used instead of multilamellar ones (20). Care must be taken, however, in preparing uniform liposome batches since smaller vesicles contain less enzyme which could result in decreased test sensitivity.

Alkaline phosphatase (AP), β-galactosidase, and horseradish peroxidase (HRP) are usually the enzyme markers of choice. These enzymes are easily monitored as are most enzyme immunoassays (EIA) (30). When AP entrapped vesicles are lysed in the presence of p-nitrophenylphosphate, this chromogenic substrate is converted into the coloured product p-nitrophenol that is spectrophotometrically determined at 410 nm (8, 22). Unlike cellular debris, liposomes having released their contents are not removed from the reaction medium since they do not cause light scattering changes. HRP-filled vesicles, when lysed, release their contents which quickly catalyze an introduced substrate with resulting oxygen consumption (20). A Clark-type oxygen probe or a similar bioselective electrode (28, 29) can be used for accurately measuring this enzymatic reaction.

Vesicles prepared with glucose, ions, or non-protein markers offer the advantage of having a protein-free and carefully controlled "ideal" liposomal model containing few possible interfering variables. These models are excellent for investigating the influence of membrane lipid composition on the interaction of immunoglobulins with liposomes. Differential glucose release, resulting from immune response and liposome damage, is related to vesicle membrane content, phospholipid fatty acyl chain length, and antibody binding affinity (25, 26). Antibodies binding to antigen-ionophore membrane conjugates can change the liposome surface so as to modulate the rate of marker ion release (8). Liposome lysis with subsequent detection of sugar markers are usually monitored spectrophotometrically while escaping ions can be measured by ion-selective electrodes or with specially built liposome immunoelectrodes (19).

A very simple, precise, and sensitive LIA method is available by using the electron spin resonance (esr) spin-labelling technique and the esr spectrometer for detecting liposome release of spin label markers (2, 24). An esr spectrum consists of broad and weak signals resulting from electron spin exchange interactions occurring among molecules confined inside liposomes. Exchange between unpaired electrons normally occurs on nitroxide radicals of spin labels such as Tempocholine chloride. Reduced signal amplitude in the esr spectrum, due to a high local concentration spin exchange, is recorded before liposome lysis. Upon marker release after lysis, however, spin label concentration and electron spin reduction results suddenly. Consequently, an immediate increase in the esr amplitude is noticed with a sharp intense signal easily detected on the esr spectrometer.

LIAs using spin labels have no radiation hazard and do not involve separation procedures. Reagent additions are not required for detecting release markers during liposome lysis. Spin label advantages include fast, simple, safe assays that require only small volumes for analysis and which can be fully automated (2). Optically opaque LIAs can be performed with spin labels while optical clarity is required when using spectrophotometrically detected markers. Multiple assays can also be performed simultaneously when using several different spin label markers. Initial cost in purchasing an esr spectrometer and training personnel is generally offset by the many advantages gained through using spin label markers.

Assay Principles

LIAs offer the advantage of rapidly measuring analytes homogeneously without requiring separation steps, several washings, or surface solid phase binding. An uninterrupted test will take place to completion in a single sample well with continued reagent additions and incubations until final results are obtained. Most LIAs are designed so that soluble antibodies bind to specific liposome surface antigens, forming immune complexes. These either precipitate upon addition of secondary antibodies or cause liposome lysis with contents being released when complement is added. Separate liposome populations loaded with different markers and surface antigens can simultaneously detect multiple antigens when present in the same analyte. However, liposomes from the same lot and lacking surface antigens can be used for measuring antibodies directly or antigens by a competitive inhibition assay. These LIAs are performed independent of complement and use antigen-cytotoxin conjugates for lysing liposomes.

Micro quantities of fluorescein loaded liposomes with surface antigens will quickly bind to specific antibodies serially diluted in microtiter plates (Fig. 3). The reaction is

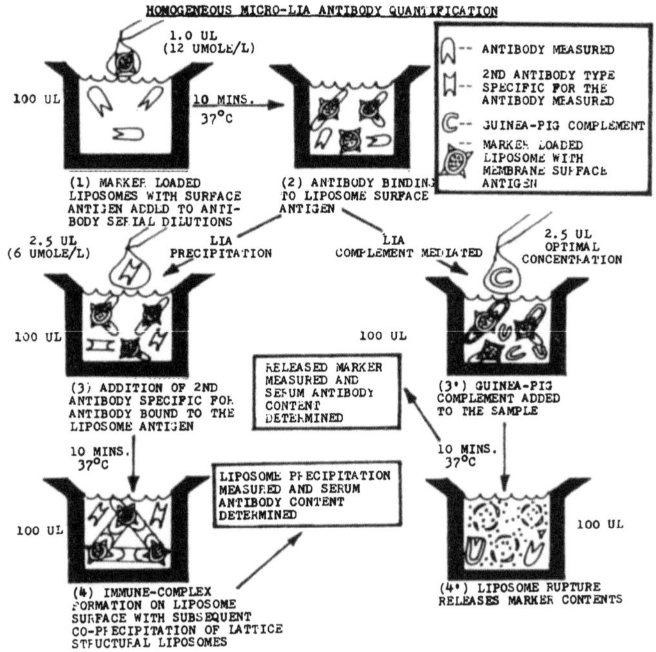

Fig. 3

usually completed within 1 hour at 20°C or 10-15 min at 37°C. Addition of secondary anti-immunoglobulin antibodies prepared against those being detected in the analyte will, when added, form a lattice structure between several immune-complexed liposomes, thus causing co-precipitation. Resulting immunoprecipitated fluorescein-labelled liposomes can be quantitated with a fluorescent microscope and the antibody content titrated. When an optimal concentration of guinea-pig complement is added instead of the secondary antibody, liposome lysis occurs with subsequent fluorescein release that is detected spectrophotometrically. Other types of dyes may be substituted for fluorescein in this type of LIA.

Complement mediated multiple antigen assays are also performed in microtiter plates. However, an optimal dilution or complement concentration must be predetermined before each assay so as to avoid non-specific liposome lysis. Usually, guinea-pig complement diluted 400 X is the optimal concentration necessary for liposome lysis in the presence of specific antibody with minimal non-specific marker release (5).

Type specific antibodies are first added to wells containing serial dilutions of an antigen mixture. After allowing to react, a precise amount of liposomes is then added. These are vesicles incorporating surface antigens identical to one of those measured in the mixture. As a result of being the same type, both the surface bound liposome and free mixture antigens will compete for specific antibodies present. After several minutes, an optimal concentration of complement is added which rapidly binds to any immune-complexes that have formed on liposome surfaces. Vesicle lysis occurs with markers released that can be measured. From this information, the amount of a specific free mixture antigen competing for certain added antibodies and which inhibits these antibodies from binding to similar liposome bound antigens can easily be calculated.

Addition of antibodies that are different from those originally present may be continued in the same microtiter wells without subsequent washing or separation steps. A new population of liposomes with surface antigens and markers that vary from those involved in detecting the first antigen are added along with complement. As before, competitive inhibition occurs and another antigen present in the sample mixture can be determined. Such a complement mediated homogeneous competitive inhibition assay (Fig. 4) which continues in the same microtiter plate without interruption is fast, economical, and ideal for automation. LIA limitations, however, include: new liposome lots with a variety of different markers and antigens required, excess costly complement along with separate specific antibodies needed to measure each antigen present, and possible error dilution factors that increase with each addition. Electron spin, incidentally, makes for excellent markers when performing LIA mul-

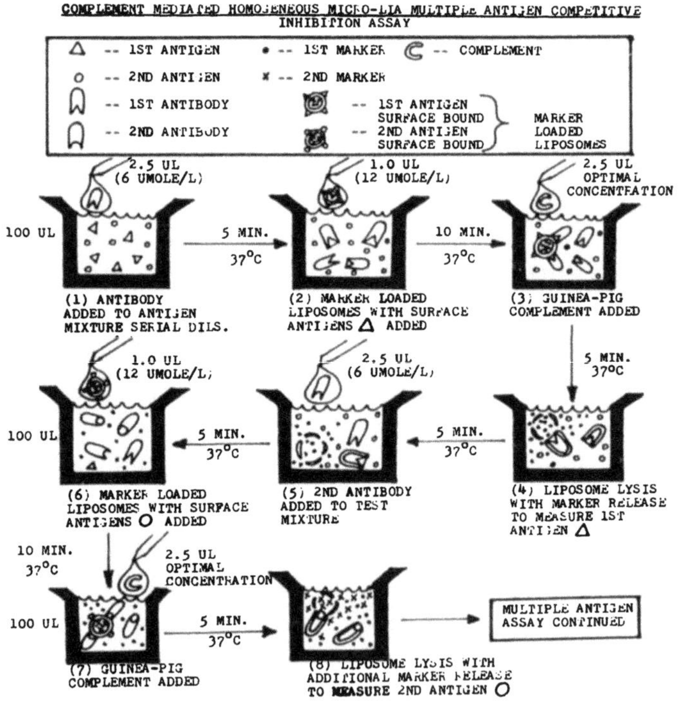

Fig. 4

tiple antigen assays. Their release from liposomes are measured directly as lysis takes place without further reagent additions.

Competitive inhibition antigen LIAs independent of complement (Figs. 5, 6) are rapidly becoming a popular and economical means of measuring multiple analytes. Single lot liposome preparations lacking surface antigens are lysed by the action of antigen-cytolysin conjugates instead of complement. Several potent cytolytic agents are available that produce lysis or liposome membrane lesions with little if any lag phase in the rapid release of markers. Such agents include gas phase singlet molecular O_2 (31), granular-lymphocyte tumour cytotoxins (32), cytolytic marine worm protein Cerebratulus lacteus toxin A-III (33), Central Asia cobra venom (27), and bee venom melittin (8, 22). These reagents and their bound antigens are required, however, to retain cytotoxic and immunologic activity after conjugation. Both homogeneous and non-competitive heterogeneous LIAs are now possible as routine assays. These offer a method with decreased potential test interferences and improved sensitivity down to nanomolar concentration range of the analyte.

Generally, less time is required during LIA incubation steps using cytolysin conjugates which quickly overcome the slow kinetics of vesicle lysis as compared with those LIAs requiring complement (Fig. 5). Homogeneous competitive inhibition antigen assays require few manipulations and can usually be completed within 10 minutes. Free antigen competes with antigen-cytolysin conjugate for a known amount of type specific antibody added. When the conjugate becomes bound to a specific antibody, however, the cytotoxic effects are quenched and liposome lysis will not occur. Probably conjugate antibody binding prevents association with liposome lipid bilayers although the exact mechanism still remains a mystery (8). Marker loaded liposomes are then introduced which immediately lyse when making contact with unbound conjugate. The amount of marker release is measured and free antigen concentration calculated.

Conversely, antibodies are detected by similar non-competitive cytolysin LIAs (Fig. 6). A direct homogeneous assay

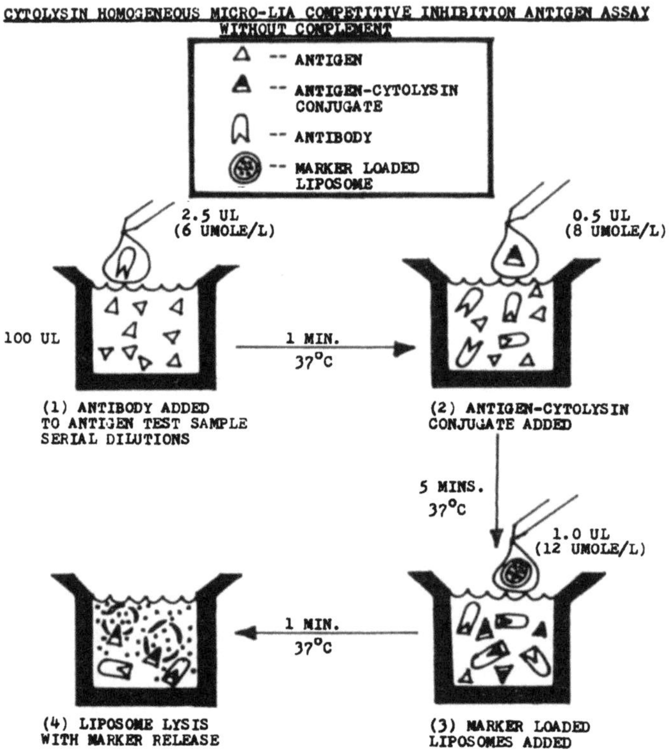

Fig. 5

requires antigen conjugate and liposome additions followed by marker release measurements. The indirect heterogeneous method involves antigen binding to a polystyrene surface. Antibodies are then added and allowed to react with fixed antigen. Washing the polystyrene microtiter wells several times prior to further additions removes unbound interfering reagents and increases test sensitivity. Conjugates, liposomes, and markers released can then be measured in a like manner as before.

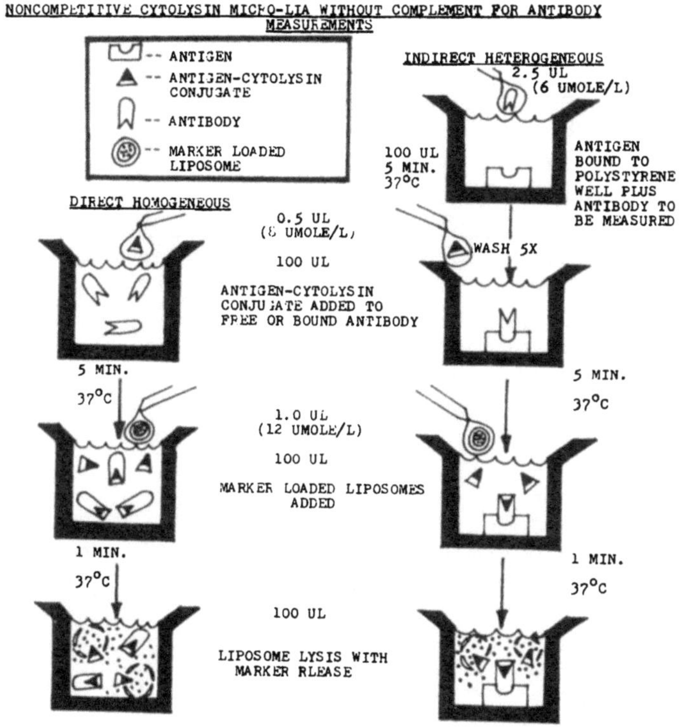

Fig. 6

References

1. Pharmacia P-L Biochemicals bulletin: Liposomes, a survey of applications. Subsidiary of Pharmacia Inc., 800 Centennial Avenue, Piscataway, N.J. 08854, 1-5 (1984).

2. Hsia, J.C., Tan, C.T.: Membrane immunoassay: Principle and application of spin membrane immunoassay. Ann. N.Y. Acad. Sci. 308, 139-148 (1978).

3. Wilson, T., Cave, L.: New commercial opportunities or liposomes emerge. Biotechnology 1, 148-154 (1983).

4. Janoff, A.S., Green, S.C., Weiner, A.L., Seibold, J., Weissman, G., Ostro, M.J.: Novel liposome composition for a rapid colorimetric test for systemic lupus erythematosus. Clin. Chem. 29, 1587-1592 (1983).

5. Ishimori, Y., Yasuda, T., Tsumita, T., Notsuki, M., Kozama, M., Tadakuma, T.: Liposome immune lysis assay (LIA): A simple method to measure anti-protein antibody using protein antigen-bearing liposomes. J. Immuno. Methods 75, 351-360 (1984).

6. Gregoriadis, G.: Liposome technology. In: "Preparation of Liposomes", Ed. Gregoriadis, G., Clinical Research Center I, 1-259 (1983).

7. Farkasrolgy, J.: A study method or producing liposomes on an industrial scale. Ann. Immunol. Hung. 23, 367-382 (1983).

8. Freytag, J.W., Litchfield, W.J.: Liposome-mediated immunoassays for small haptens (digoxin) independent of complement. J. Immuno. Methods 70, 133-140 (1984).

9. Bangham, A.D., Standish, M.M., Watkins, J.C.: Diffusion of univalent ions across the lamellae of swollen phospholipids. J. Mol. Biol. 13, 238 (1965).

10. Deamer, D.W., Bangham, A.D.: Large volume liposomes by an ether vaporization method. Biochim. biophys. Acta 443, 629 (1976).

11. Schieren, H., Rudolph, S., Finkelstein, M., Coleman, P., Weissman, G.: Comparison of large unilamellar vesicles prepared by a petroleum ether vaporization method with multilamellar vesicles: ESR, diffusion and entrapment analyses. Biochim. biophys. Acta 542, 137 (1978).

12. Papahadjopoulos, D., Vail, W.J., Jacobson, K., Paste, G.: Cochleate lipid cylinder formation by fusion of unilamellar lipid vesicles. Biochim. biophys. Acta 394, 483 (1975).

13. Szoka, F., Papahadjopoulos, D.: Procedure for preparation of liposomes with large internal aqueous space and high capture by reverse-phase evaporation. Proc. natn. Acad. Sci. U.S.A. 75, 4194 (1978).

14. Papahadjopoulos, D., Miller, N.: Phospholipid model membranes. I. Structure characteristics of hydrated liquid crystals. Biochim. biophys. Acta 135, 624 (1967).

15. Huang, C.H.: Studies of phosphatidylcholine vesicles: Formation and physical characteristics. Biochemistry 8, 344 (1969).

16. Batzri, S., Korn, E.D.: Single bilayer liposomes without sonication. Biochim. biophys. Acta 298, 1015 (1973).

17. Milsmann, M.H.W., Schwendener, R.A., Weder, H.G.: The preparation of large single bilayer liposomes by a fast and controlled dialysis. Biochim. biophys. Acta 512, 147 (1978).

18. Battelle Memorial Institute, British Patent Application 2001929A (1979).

19. Shiba, K., Watanabe, T., Umezawa, Y., Tujiwara, S.: Liposome immunoelectrode. Chemistry Letters 2, 155-158 (1980).

20. Haga, M., Sugawara, S., Itagaki, H.: Drug sensor: Liposome immunosensor for theophylline. Anal. Biochem. 118, 286-293 (1981).

21. Jansson, B., Sjögren, H.O.: Inhibition of antibody binding to viable cells by liposomes containing purified rat spleen glycoprotein. J. Immuno. Methods 65, 159-169 (1983).

22. Litchfield, W.J., Freytag, J.W., Adamich, M.: Highly sensitive immunoassays based on use of liposomes without complement. Clin. Chem. 30, 1441-1445 (1984).

23. Axelsson, B., Ericksson, H., Borrebaeck, C., Mattiasson, B., Sjögren, H.O.: Liposome immune assay (LIA). Use of membrane antigens inserted into labelled lipid vesicles as targets in immune assays. J. Immuno. Methods 41, 351-363 (1981).

24. Vistnes, A.I.: A new method of evaluating complement mediated lysis of liposomes. J. Immuno. Methods 68, 251-261(1984).

25. Haxby, J.A., Kinsky, C.B., Kinsky, S.C.: Immune response of a liposomal model membrane. Proc. natn. Acad. Sci. U.S.A. 61, 300-307 (1968).

26. Alving, C.R., Joseph, K.C., Wistar, R.: Influence of membrane composition on the interaction of human monoclonal "anti-Forssman" immunoglobulin with liposomes. Biochemistry 13, 4818-4824 (1974).

27. Aripov, T.F., Salakhutdinov, B.A., Salikhova, Z.T., Sadykov, A.S., Tashmukhamedov, B.A.: Structural changes of liposomes phospholipid packing induced by cytotoxin of the central Asia cobra venom. Gen. Physiol. Biophys. 3, 489-496 (1984).

28. Monroe, D.: Bioselective electrodes, part one. American Clinical Products Review 4, 24-34 (1985).

29. Monroe, D.: Bioselective electrodes, part two. American Clinical Products Review 4, 46-57 (1985).

30. Monroe, D.: Enzyme immunoassay. Anal. Chem. 56, 921A-931A (1984).

31. Eisenberg, W.C., Taylor, K., Grossweiner, L.I.: Lysis of egg phosphatidylcholine liposomes by singlet oxygen generated in the gas phase. Photochem. and Photobio. 40, 55-58 (1984).

32. Blumenthal, R., Millard, P.J., Henkart, M.P., Reynolds, C.W., Henkart, P.A.: Liposomes as targets for granule cytolysin from cytotoxic large granular lymphocyte tumors. Proc. natn. Acad. Sci. U.S.A. 81, 5551-5555 (1984).

33. Blumenthal, K.M.: Release of liposomal markers by Cerebratulus Toxin A-III. Biochem. biophys. Res. Commun. 121, 14-18 (1984).

DETERGENT SOLUBILISED ANTIGENS IN ENZYME IMMUNOASSAY WITH PARTICULAR REFERENCE TO ENZYME-LINKED IMMUNOSORBENT ASSAY (ELISA) SYSTEMS

O. L. Goldring
Department of Biological Sciences, North East Surrey College of Technology, Reigate Road, Ewell, Epsom, Surrey, KT17 3DS, U.K.

Introduction

Despite numerous publications on the subject, it must be remembered that enzyme immunoassay (EIA) is a relatively new technique compared to radioimmunoassay (RIA), and some aspects of the methodology of EIA still require detailed evaluation (1).

EIA is widely used in medical immunodiagnosis (2), and has become popular as a technique for screening culture supernatants during monoclonal antibody (MCAB) production (3). EIA is used as a procedure to demonstrate the antigenicity of purified or partially purified protein preparations (4), and is also used to screen for drugs/pharmacological agents, in body fluids (5).

Most of the literature on enzyme-linked immunosorbent assay (ELISA) (either "indirect antibody" or "sandwich" antigen systems, 1), reports the use of water soluble (as opposed to detergent soluble) forms of antigen in the assay procedures. Sometimes whole organisms (i.e. protozoal parasites) are used as antigens. On the other hand, there has been little use of detergent extracted antigens in EIA, and the literature on the topic is comparatively sparse. This is surprising because there must exist a great many membrane bound protein antigens which, were they soluble, would be of potential interest and use in EIA systems.

To-date, the use of detergent extracted antigens in bacterial, fungal, and parasitic ELISA systems has not been a prominent feature of the literature on ELISA. The literature on detergent extracted antigens in virological ELISA systems is more extensive, but not profuse. There are reports of the use of detergent extracted antigens in ELISA for biochemical and immunological studies, where the main emphasis of the work has been on characterisation of membrane proteins (and their antigenicity), but the literature is not extensive.

This article gives a brief summary of detergent action on membrane systems and then draws attention to the use of detergent solubilised antigen systems in EIA. In particular, attention is given to the three areas listed below.

What range of detergents are available and which of these are suitable for use in ELISA ?

What methods of solubilisation appear to be most effective?

What problems occur concerning the general use of detergents during ELISA ?

1. Detergent Action on Membrane Systems

The proteins of biological membranes can be categorised as either 'peripheral' or 'integral' ('intrinsic') proteins (6). This is based on the way these proteins can be detached from the membrane. Peripheral proteins are usually water soluble after extraction. Integral proteins are tightly membrane bound and can only be solubilised by disrupting the membrane with organic solvents or by using detergents.

Detergents have been defined as "soluble amphiphiles that effectively bring about the solubilisation of membrane lipid" (6). In aqueous solutions, detergent molecules occur in the form of monomers and micelles. At low concentration, detergents exist as monomers only, whilst, at high concentration, both monomers and micelles exist in equilibrium. Micelles are defined as "thermodynamically stable colloidal aggregates, spontaneously formed by amphiphiles above a narrow concentration range (the critical micellar concentration or CMC), at temperatures above the critical micellar temperature". The CMC also represents the highest detergent chemical potential (7). In a micelle, the apolar groups of the detergent molecules are directed inwards, whilst the polar groups are directed outwards. (Micelle size increases with increasing size of apolar moieties).

The precise physical chemistry of membrane solubilisation is not altogether certain (6), and it is not considered appropriate to attempt to cover this aspect in detail in the present paper. However, once solubilised by detergents, lipids and membrane proteins can become incorporated into the detergent micelles, with the resulting formation of mixed micelles (6, 7).

With increasing amount of detergent, membranes undergo different states of disorganisation starting with lysis or rupture (7), The selective extraction of membrane proteins occurs at around 0.1-1.0 mg detergent/mg membrane lipid (7). The membrane bilayer remains intact. When detergent concentration increases (2 mg detergent/mg membrane lipid) the original lamellar structure may disappear and the components of the membrane become soluble (6, 7). It has been reported that 10 mg of detergent (or more)/mg membrane lipid (i.e. excess detergent) is often needed to render integral membrane proteins lipid free (i.e. to delipidate them), and to incorporate the protein into a water soluble micelle with the detergent (7).

Therefore, ultimately, one can have mixed micelles of the following type: (a) lipid-protein-detergent; (b) protein-detergent; (c) lipid-detergent. It is also reported (7) that, if chromatographic procedures are to be used, then to ensure the separation (i.e. chromatographically) of detergent solubilised proteins, the 'solvent' requires a detergent concentration of the order of 1.5-2 micelles/protein molecule, because, otherwise, proteins which could be separated may be trapped in the same micelle.

To summarise, the membrane proteins have been solubilised by detergent during the initial extraction process and, subsequently, they exist as part of a water soluble complex consisting of the three types of micelles described above (a, b, c).

This initial extract can be manipulated by many of the techniques of protein chemistry (8), and, to a limited extent, by analysis using quantitative immunological methods, including enzyme immunoassays. In the context of EIA it is clear that few workers use as "antigen" anything other than "the initial extraction complex", i.e. the whole soluble complex consisting of the type (a), (b) (c) micelles, described above.

2. Types of Detergents used in ELISA

The number of different detergents available is large, and many have been used in biochemical studies. For the purposes of this article, detergents will be placed into four groups:

A. Anionic detergents (6) (which will also include sodium deoxycholate, a bile salt, 9, 10)
B. Cationic detergents (6, 7, 9)
C. Zwitterionic detergents (7, 9)
D. Nonionic detergents (6, 7, 9).

Examples are given in each group, including chemical or trade names (for detailed structural chemistry and properties, other publications are available, 6, 7, 9, 10).

Examples are given of the use of each detergent in ELISA (or related immunological assays). The prime use of the detergent was either in solubilising antigen or studying detergent influence on the reaction kinetics of the assay.

A. Anionic detergents

These include sodium dodecyl sulphate (SDS), sodium dodecyl-N-sarcosinate, sodium dodecyl sulphonate (6), and sodium deoxycholate (7, 9, 10).

(1) Sodium dodecyl sulphate (SDS). SDS despite its ability to solubilise proteins from whole cell extracts, tends to cause precipitation of artifacts, or antigen (ag)/antibody (ab) dissociation (11). At SDS concentrations greater than 2%, this

certainly occurs in agarose gels. Possibly because of extrapolation from these types of observations, SDS has not been used to any great extent in EIA.

However, SDS has been used in similar systems, such as radioimmunoassay, to study link proteins from nasal cartilage (12). SDS has also been cited as a possible denaturing (de-aggregating) agent with which to inhibit antigen aggregation, a problem which occurred during EIA of link protein (13).

There is at least one report of an ELISA study concerning SDS (14). Influenza A/FM/1/47 virus was attached to microtitre trays. Subsequent treatment of the virus with 0.025% SDS was then found to change the accessibility of the haemagglutinin determinant sites on the viral surface for subsequent antibody binding. Of all detergents used, SDS was the most inhibitory in this respect.

It is possible, however, that in this instance SDS was actually removing the haemagglutinin off the viral surface. It is interesting to note that when the haemagglutinin was first isolated from the viral surface, was then attached to the microtitre plastic surface, and was subsequently treated with 0.025% SDS, the reactivity of the haemagglutinin for antibody was not changed. SDS did not therefore affect isolated haemagglutinin, only haemagglutinin in its native state attached to the virus.

It is also worth recording that, despite its inherent problems of causing protein denaturation, SDS is successfully used in the technique of combined SDS-polyacrylamide gel electrophoresis and immunoprobing ("blotting") using nitrocellulose membrane bound antigens (15).

Perhaps by sequestering non-protein bound SDS in mixed micelles, using Triton X100 after SDS sample solubilisation (11), SDS may be utilisable in EIA studies. Alternatively, SDS may be removed by other methods (8).

(2) Sodium deoxycholate (DOC). This is a member of the 'bile salts' family of detergents. Some authors prefer to treat the bile salts as a separate group of detergents, distinct from

other ionic detergents. This distinction is based on their rigid steroid ring structure and different detergent properties (7). For the purpose of this article, however, DOC will be included in the ionic series of detergents as an example of an anionic detergent (9, 10).

DOC has been used in a number of ELISA studies (14, 16, 17). It has been employed to solubilise components of the outer membranes of Brucella ovis (17), the cell membrane of herpes simplex type 1 infected (HSVI) BHK-21 cells (16), and has also been used to study the binding of monoclonal antibody (MCAB) to influenza haemagglutinin (14).

The study of B. ovis (17) is a good example of the use of detergents in EIA. The study examined which antigenic preparation out of three different types of preparation of B. ovis (including a detergent extracted antigen), was the most suitable for assessing antibody levels in sera obtained from three different groups of animals (previously vaccinated, or experimentally infected, or naturally infected with B. ovis).

When whole intact B. ovis was used as antigen for a titration profile of serum from vaccinated rams (IM-S), experimentally infected (IV-S), and naturally infected rams (InF-S), the whole bacteria successfully bound antibodies present in all three categories of ram sera. An LPS heat-stable component only reacted with IM-S, whereas a DOC extract of B. ovis outer membranes (which had then been chromatographed on Sephacryl 200) was shown to have an active fraction (Peak B) which reacted preferentially with InF-S. Peak B antigen (DOC extract) could detect an activity in InF-S diluted up to 1 : 6400. This study demonstrated clearly that the choice of antigen may determine the subsequent usefulness of an ELISA test system.

DOC was also successfully used to solubilise HSVI infected BHK-21 cell membranes (16). A DOC concentration of 7 mM which was just above the critical micellar concentration (CMC) of 4-6 mM (9) gave almost complete membrane solubilisation. Significantly, this paper (16) also demonstrated that DOC concentration of 100 mM (4.1% w/v) did not interfere with adsorption of antigen to the solid phase. This was in distinct contrast to the use of the nonionic detergent Triton X-100 (TX 100) where adsorption of membrane associated antigens to the solid phase was reported to be inhibited, except at a detergent concentration of 0.001% or 0.0001%, which is well below the CMC (0.01%) of Triton X-100 (16).

DOC (compared to TX 100) did not apparently inhibit the monoclonal antibody binding capacity of the haemagglutinin of influenza A/FM/1/47 virus (14). However, this is not the only possible explanation for the DOC observations. It is possible that, in this particular instance, DOC was simply not effective in solubilising the haemagglutinin from the viral surface, whereas TX 100 was effective.

In conclusion, of the available anionic detergents, as for example sodium dodecyl sulphate, sodium dodecyl-N-sarcosinate (for which there appears to be no reported literature in EIA), and sodium deoxycholate, the latter would seem to offer most promise in EIA. It is surprising that DOC does not appear to have been as widely used in EIA as has TX 100 or Nonidet P40 (NP40).

B. Cationic detergents

Three cationic detergents are commonly listed (6): Cetyltrimethylammonium bromide (CTAB), tetradecylammonium bromide, and dodecylpyridinium chloride. Of these only CTAB appears to have been studied in EIA systems, and even then not directly as a detergent for solubilising membrane antigens.

(1) Cetyltrimethylammonium bromide (CTAB). The use of CTAB is reported in ELISA (14), in the study concerned with the haemagglutinin of influenza A/FM/1/47 virus, to which previous reference has been made in connection with both SDS and DOC.

The isolated haemagglutinin was bound to microtitre plates, treated with various detergents including CTAB, and then assessed for its ability to bind MCAB. CTAB (as well as DOC, SDS, and octyl glucoside) improved the MCAB binding capacity of the antigen by 20%, when compared with the antigen pretreated with Tween 20. Therefore when in an isolated form, the haemagglutinin MCAB binding capacity was influenced (improved) by exposure to CTAB. This was not the case when the virus was pre-adsorbed onto the microtitre trays and then treated with CTAB. As before (with both SDS and DOC), the MCAB binding was considerably reduced. Again, this may reflect removal of the haemagglutinin from the virus and not inhibition of MCAB binding (this has been mentioned earlier with SDS and DOC), in which case CTAB could be considered an effective detergent for solubilising the haemagglutinin. In this context it is worth noting, however, that DOC and octyl glucoside (OG) were not effective in altering the MCAB binding ability of the haemagglutinin when virus was first pre-adsorbed to the microtitre tray, and then treated with DOC or OG.

Irrespective of whatever was occurring, either haemagglutinin solubilisation (release), or changed accessibility of epitope determinants, the study (14) illustrates a practical concept which has been made elsewhere (7), that no one detergent can be singled out as "most effective" for membrane solubilisation. The optimal detergent for a particular membrane or membrane protein frequently has to be found empirically,and the choice may depend on the type of experiment one wants to carry out (7).

C. Zwitterionic detergents

The use of this class of detergents in EIA is apparently extremely limited. In fact, of all types of detergents, anionic, cationic, nonionic, and zwitterionic, the latter category seems to have received least attention of all in EIA. This is surprising because, in other biochemical studies, zwitterionic detergents are reported to be particularly effective at disaggregating hydrophobic proteins (18). It has been suggested that lysophospholipids could be expected to be especially mild in their effects on sensitive membrane proteins (although it was stated at the time that not enough experimental information was available to provide definite support for this concept) (7).

Three main categories of zwitterionic detergents are available: (1) The Zwittergents (including 3:12, 3:14 and 3:16); (2) CHAPS:3-[(3-cholamidopropyl) dimethylammonio]-1-propane sulphonate; (3) lysophospholipids, including synthetic lysophospholipids.

(1) Zwittergent 3-12: Zwittergent 3-14 etc. (Zwittergent is a trademark of Cal/Biochem/Behring Diagnostics Inc.). The binding of the 'link protein' component of proteoglycan aggregate to polystyrene plates in ELISA was inhibited by Zwittergent 3-12 (Z3-12) (13). Inhibition of binding also occurred with TX 100 and Tween 20. However, no mention was made in the paper (13) of any ability of Z3-12 to solubilise 'link protein', and at the time of writing the current article, there does not appear to have been any extensive use of the Zwittergents in EIA.

(2) CHAPS. Although apparently unused in EIA, CHAPS has been reported to exhibit superior membrane protein solubilising characteristics to nonionic detergents. Replacement of Nonidet P40 (NP40) with CHAPS in isoelectric focusing of rainbow trout liver microsomes, increased resolution considerably in a two-dimensional electrophoretic system, and was also reported to eliminate much of the background gel smearing sometimes associated with other detergents (TX 100/NP40) (18).

(3) Lysophospholipids. Lysophospholipids result from the action of phospholipases on diacyl phospholipids, and small amounts of lysophospholipids are present in most membranes. When added to cells these zwitterionic detergents cause lysis and cell fusion, and at higher concentrations can solubilise membranes (7).

The use of lysophospholipids in EIA apparently remains to be documented, but because of their mild action, these detergents may prove suitable. This requires further study.

D. Nonionic detergents

This has been the most extensively used group of detergents in EIA studies.

In contrast to many of the ionic detergents which can frequently cause protein denaturation (7), nonionic detergents with polyoxyethylene or sugar head groups do not usually denature proteins. However, although efficient in breaking lipid-lipid and lipid-protein interaction in membranes, nonionic detergents are reported to be ineffective in breaking interactions between proteins (7). Also included in the nonionic detergents (but which are not polyoxyethylene based) are the B-D-alkyl glucosides.

There are excellent descriptions of the variety of polyoxyethylene detergents (6, 7). For the purpose of the current article, these polyoxyethylene detergents will be placed into three groups (7).

Group 1: Alkylpolyoxyethylene Ethers: which include the BRIJ series (Atlas Chemical Industries Inc/and Sigma Chemical Co. Ltd.), the Emulphogene series (GAF Corp NY), and the Lubrol series PX (Lubrol 12 A9:ICI) and WX (Lubrol 17 A17:ICI).

Group 2: Alkylphenylpolyoxyethylene Ethers: which include the Triton X series (Rohm and Hass Co. Philadelphia), Nonidet P40 (Gallard Schlesinger Chemical Mfg. Co. NY), the Tergitol series (Union Carbide Corp. NY), and the Igepal series (GAF Corp. NY).

Group 3: Acylpolyoxyethylene Sorbitan Esters: which include the Tween series (Atlas Chemical Industries Inc.), and the Emasol series.

(1) Alkylpolyoxyethylene ethers (group 1). Lubrol WX (in conjunction with TX100) has been used to isolate 'thermogenin' (the brown adipose tissue-specific, GDP binding, 32kDa uncoupling protein) from the breakthrough fraction of the detergent extract applied to hydroxylapatite columns (19). 'Thermogenin' was then identified in a competitive ELISA assay, and also by immuno-dot blotting on nitrocellulose membranes.

(2) Alkylphenylpolyoxyethylene ethers (group 2). Of this series only two detergents have received much attention in EIA, these being Triton X100 and Nonidet P40. In fact, these have been the most extensively used of all detergents in EIA, particularly in ELISA. Chemically, Triton X100 and Nonidet P40 are very similar (7), but for the purposes of this article will be considered separately.

Triton X100 (TX 100). An early report on the use of TX 100 which, although it was not an EIA study, indicated that TX 100 was very efficient in isolating two surface glycoproteins and a non-glycosylated membrane protein from Newcastle disease virus (NDV). This isolation was done using 2% TX 100 and IM KCl (20).

Following this same solubilisation procedure, a study ten years later (21) used monoclonal antibodies (MCAB) against the HN glycoprotein of D_{26} strain of NDV in an ELISA assay. Purified virions were disrupted with 2% TX 100 and IM KCl, were then diluted in standard carbonate/bicarbonate coating buffer, and were used as antigen. The study showed that the HN glycoprotein of NDV had at least three distinct domains which were distinguishable with the MCAB.

A similar study was carried out using Sindbis virus (SIN) disrupted with TX 100. This antigen was then used in a competitive ELISA using MCAB to SIN (22). Apart from its efficient application as a detergent for disrupting SIN, TX 100 was also useful in this particular study because it revealed an important additional function of detergent treatment which is their ability to expose otherwise cryptic (inaccessible to antibodies) epitopes on the surfaces of infected virions. This study was also a useful example of a situation where TX 100 had no inhibitory effect in EIA. This has not always been the case (13, 16).

In contrast, pretreatment of the haemagglutinin on the viral surface of influenza A/FM/1/47 virus with a small (0.025-0.05%) amount of TX 100 was reported (14) to strongly change (inhibit) the accessibility of the haemagglutinin determinant sites for antibody binding (but TX 100 did not alter the binding properties of the isolated haemagglutinin pre-attached to the solid phase surface). This example has been mentioned before for SDS, DOC, and CTAB. The effect could have been due to haemagglutinin release from the viral surface caused by the detergent action of TX 100 (i.e. actual physical loss of the haemagglutinin due to subsequent washing away from the solid phase surface), and not an effect apparently due to the inhibitory action of TX 100. Therefore, this may represent a situation where the particular assay system used was inappropriate and was not a case of the detergent (TX 100, DOC, SDS, or CTAB) causing inhibition of antibody binding.

A good example for the use of detergents in EIA is seen in the development (23) of an indirect solid phase ELISA for the detection of herpes simplex virus (HSV) antigens in clinical specimens. Microtitre plates were coated with guinea-pig anti-herpes simplex Type 1 immunoglobulins. Clinical specimens were diluted in phosphate buffered saline (PBS) containing 20% foetal calf serum (FCS), 2% Tween 20, and 1% TX 100, and incubated on the plates overnight at 37°C. This HSV antigen was detected using rabbit anti-HSV immunoglobulins. The test was specific, and gave a 74% success rate (which equalled RIA).

Nonidet P40 (NP40). Bovine coronavirus infections are characterised by severe enteritis, and ELISA has been used for sero-epidemiological studies (24). Virus was grown in Madin-Darby bovine kidney cells. Two techniques were used, ELISA and passive haemagglutination (PHA), to evaluate antibody levels in experimentally infected gnotobiotic calves. Three types of antigen were used, purified coronavirus (CV), virus infected Madin-Darby bovine kidney cells (VIMD), and a 0.5% NP40 extract of purified coronavirus (NP40CV).

CV and NP40CV were suitable as antigen in ELISA. It must be stated however that, although PHA and ELISA were of similar specificity and sensitivity, PHA allowed 300 times more tests to be performed per volume of virus infected cell culture than did ELISA.

In a completely different field, ELISA has been used to

identify and quantitate T lymphocyte (T cell) antigen(s) from bovine thymocytes and bovine peripheral blood T cells. ELISA used 0.5% NP40 soluble membrane extracts of T cells as antigen (25).

The time dependent loss of sperm maturation antigen 4(SMA4) from mouse sperm has been examined by ELISA using NP40 extracted sperm (26).

(3) Acylpolyoxyethylene sorbitan esters (group 3). These include the Tween series (Tween 21, 81, 20, 40, 60, 80) and Emasol. There are apparently no reports of the use of Emasol in EIA. Of the Tween series, Tween 20 (polyoxyethylene sorbitan monolaurate) is most extensively used. Its main application, however, is as a component of the ELISA 'washing' or 'diluting' fluid, and not as a detergent with which to extract antigens.

Tween 20 is generally found useful when included in the ELISA washing buffer. This was exemplified by a rubella EIA system in which the omission of Tween 20 from the washing buffer produced erratic results, in that otherwise negative or low titre sera, occasionally gave high absorbances (27).

In another study, Tween 20 (at 0.05% dilution) also decreased the binding of an interfering non-viral EIA active material found in stool extracts of individuals who were secreting hepatitis A virus (28).

Tween 20 also exerts an effect on one of the various chromogens which are used in horseradish peroxidase (HRP) EIA systems. HRP-IgG conjugate was bound to the solid phase and detected using various substrates with or without 0.05% Tween 20. The substrates used were ABTS, o-phenylenediamine (OPD), o-dianisidine (OD), and 4-aminoantipyrene (4AAP). Absorbance values were not increased when ABTS, OPD or 4AAP contained 0.05% Tween 20. However, application of OD as chromogen in 0.05% Tween 20, gave a 1.7 times higher absorbance after 60 minutes, and a 2.0 times higher absorbance after 120 minutes of reaction, than absorbance values for comparable times using OD without Tween 20 (29).

(4) B-D-Alkyl glucosides. A category of detergents which are recent additions to the nonionic family of detergents, are the B-D-Alkyl glucosides (7). They appear to be at least as gentle as the polyoxyethylene detergents, but only the octyl derivative has received much attention (7). The octyl derivative, octyl-B-D-glucopyranoside has an exceptionally high critical micellar concentration of 25 mM (0.74%).

Octyl-B-D-Glucopyranoside (OG). Little is reported in the EIA literature about the use of OG. It has been found, however, (14) that OG either (a) was unable to solubilise the haemagglutinin on the surface of influenza A/FM/1/47 virus or (b) was unable to alter the accessibility of the relevant antibody binding determinants of the haemagglutinin on the viral surface as measured by ELISA. Nevertheless, when the haemagglutinin was pre-isolated, attached to the plastic surface, and then treated with OG, this detergent then markedly increased the ability of MCAB to bind to the haemagglutinin in an ELISA system. This same study, previously mentioned, also investigated SDS, DOC, CTAB and TX 100.

OG has also been used to solubilise mouse MHC Type II antigens from spleen cells (30). These antigens were then incorporated into artificial liposomes which were subsequently used to present antigen to cloned T cells and T cell hybridomas. Although this study did not involve EIA, the conditions of use of OG in the above system could be relevant to future EIA investigations in which OG is used.

3. Methods for Antigen Solubilisation with Detergents in ELISA Systems

It is not always possible to state what amount of detergent is "necessary" to extract all the protein out of a biological membrane (31). This is because different membranes may have differing amounts of protein and lipid (32), and it is ultimately the ratio of detergent:lipid (w/w) which determines the degree of solubilisation of the membrane (6).

The ratio of lipid:protein may vary between extremes of 3:1 for myelin to 1:3 for mitochondria - hence, some membranes may have 25% of their total weight as lipid, whilst others may have 75% of their total weight as lipid (32).

It has been reported in many non-EIA systems that 2 mg of detergent:1 mg membrane lipid is an adequate ratio to allow membrane solubilisation to occur, with a resulting complex of protein-lipid-detergent micelles, lipid-detergent micelles, and protein-detergent micelles. This would favour the formation of the first two types of micelles (7). However, in order to ensure that the bulk of the micelles occur as protein-detergent type, i.e. in order to delipidate the protein-lipid-detergent micelles, it would be necessary to have detergent in excess, at ratios of 10 mg detergent:1 mg lipid (7). Nevertheless, it has been reported that many proteins can lose their biological activity in the complete absence of lipid (31).

The same critical ratio of 2 mg detergent:1 mg membrane lipid may well apply when preparing detergent treated membrane antigens for solid phase EIA, but there is little data on which to base a retrospective analysis, i.e. to calculate what ratios have actually been used in detergent EIA.

For most of the detergent EIA literature previously quoted, it is not possible to state with certainty that the detergent was used in conditions of excess during antigen extraction. If the detergent was in excess, then it is likely that complete membrane solubilisation would have resulted. Given the high concentrations of detergents used, it is probable (but not proven), that detergent was in excess in most instances. This assumption does require verification however. Other non-EIA observations have indicated that a general detergent:lipid ratio (w/w) required for membrane solubilisation with TX 100 was 1.9 $\pm$ 0.9 SD; for SDS was 1.6 $\pm$ 0.3 SD, and for DOC was 1.6 $\pm$ 1.35 SD (6).

In practice very few EIA workers will determine the detergent:lipid ratio. This is understandable, but on the other hand, if it is considered desirable to be able to reproduce the results of solubilisation, then the concentration of the membrane as well as that of detergent should be reported. It is known that the effects of detergents correlate well with the total detergent:membrane ratio in the sample. The lower the CMC of the detergent, and the higher the concentration of the membrane, then the closer will be the total detergent:membrane ratio to the bound detergent:membrane ratio (6). A membrane concentration (or whole cell, or organism concentration) reported in mg protein/ml, or mg/ml, if not mg lipid/ml, would at least ensure that anyone who is subsequently interested in repeating a particular experiment, has a reasonable likelihood of being able to standardise the solubilisation procedure initially used.

A further point to consider is the end point of solubilisation. This can be taken as the ratio of detergent:membrane at which no significant additional release of membrane material is observed upon further addition of detergent (6). It would be interesting to know in what proportion of the EIA detergent literature the end point of solubilisation has been determined because this type of data does not appear in the published results.

The minimum concentration usually required for a detergent to commence lysis or solubilisation (9) (not necessarily optimal solubilisation,6, 7, 9) of a membrane will be the CMC, i.e. detergent micelles must be present.

In most of the literature on EIA where final detergent concentrations are mentioned, the figures range between 0.5-2.0% (5 mg/ml-20 mg/ml). As has been indicated, rarely is mention made of the membrane concentration (mg/ml, mg protein/ml, mg lipid/ml). It is probable therefore that the amount of detergent used in most of the detergent EIA literature does not represent an absolute value requirement for either maximum membrane solubilisation to occur, or for maximum antigen binding to occur to the solid phase (at minimum detergent concentration). In the case of antigen binding, the reverse may in fact be true i.e. a lower initial concentration of detergent (but bearing in mind the previous statements about CMC and detergent:membrane ratio), or subsequent detergent removal from antigen, may facilitate greater binding of antigen to the solid phase.

In general terms, many of these facets of the assays have not been investigated for most of the detergent EIA systems described. It is worth remembering that there are two major problems to consider when preparing detergent treated antigens for solid phase assays. These are:

(1) to obtain a high yield of solubilised antigen,

(2) to retain the antigen affinity for the solid phase (or the affinity for the coating antibody if the test is of a 'sandwich' antigen type).

These two aspects are not always compatible (16), and when developing any new detergent EIA, consideration should be given to the parameters outlined in the preceding paragraphs of this section.

Taking the above statements into account, no attempt will be made in the remainder of this section to indicate an optimum concentration for any particular detergent. General outlines will be briefly given instead, of the methods of detergent solubilisation already in use in ELISA. These will give some idea of the range of detergent concentrations which have been used in detergent EIA. The proviso is made that this methodology will only apply to the detergents most frequently used in ELISA, these being sodium deoxycholate (DOC), Triton X 100 (TX 100), and Nonidet P40 (NP40).

Sodium deoxycholate (DOC)

The ionic detergent which has been most widely studied in ELISA is sodium deoxycholate (DOC). DOC has been applied to both viral (14, 16) and bacterial (17) ELISA systems.

In one of the viral ELISA systems (14), DOC was used for 1 hr at 0.050% in PBS (pH 7.2-7.4) in an attempt to extract vital haemagglutinin. The results indicated that DOC was unable to facilitate extraction of the haemagglutinin under these conditions. In the other viral ELISA study (16), the final concentration of DOC used was approximately 0.3%, and the final extraction buffer was 0.01M glycine-sodium hydroxide (pH 8.8). Detergent extraction was done on ice, and the initial antigen was a membrane preparation of virally infected BHK-21 cells. The final detergent extracted antigen was a 100,000 x g supernatant (1 hr) of the DOC treated virally infected BHK-21 cell membranes. This antigen worked well in the assay system with a coating buffer of 0.05M Tris-HCl (pH 8.2). It is worth noting that there was no interference of antigen binding to the solid phase when further DOC was added to the antigen containing 0.05M Tris-HCl buffer, so as to make the final DOC concentration 100mM (4.1% w/v).

DOC was also used to solubilise components of the outer membranes of Brucella ovis (17). Extraction was performed using 50 mM glycine, 10 mM Na_2 EDTA, and 1% DOC, pH 9.5. This study (17) was also one of the few to separate the antigen further, and used Sephacryl 200.

These publications indicate that a range of concentration of DOC (0.05-1.0%) has been used in ELISA systems, but as previously stated, an optimal concentration should be determined for any new investigation.

Finally, mention should be made of the necessity of a high pH buffer and adequate ionic strength for DOC action to be most effective (6, 7, 31). This will be discussed later in section 4 dealing with problems arising during detergent EIA work.

Triton X 100 (TX 100)

The methodology for use of TX 100 is adequately illustrated in a number of investigations (21, 22, 23). These are all applications in the use of ELISA applied to viral research.

In one study, Newcastle disease virus (NDV) was disrupted with 2% TX 100 and 1M KCl (21), and then simply diluted in coating buffer pH 9.6 to a concentration of 100 µg protein/ml. This solution was then used as antigen. Adherence to the plastic was allowed to take place at 4°C overnight.

This is the simplest procedure described, and no problems resulted. It would be interesting to know what was the final concentration of TX 100 in each well during antigen coating, because it has been reported, in another system, that a concentration of TX 100 greater than 0.001% was inhibitory, and blocked Herpes simplex viral antigen adsorption to the solid phase (16).

In another study, using Sindbis virus (22), the virus was disrupted with 1.0% TX 100, and was then diluted in phosphate buffered saline (PBS), and used as antigen for coupling to the solid phase. Again, no problems occurred with this simple procedure.

An antigen ELISA, detecting Herpes simplex virus antigens in clinical specimens, disrupted the virus in the specimens using a PBS/20% FCS, 2% Tween 20, 1% TX 100, and merthiolate 40 µg/ml, buffer (23).

TX 100 is obviously effective at disrupting virus, and the antigens can be assayed with little difficulty. What about other types of sample, i.e. bacteria, parasites, tissue proteins, etc? The application of TX 100 to other systems is discussed later in this section.

Nonidet P40 (NP 40)

NP40 is chemically similar to TX 100 (6, 7, 9). It is also one of the most popular detergents in EIA and has been applied as a detergent for solubilising as widely differing structures as bovine enteric coronavirus (24), bovine thymocyte and peripheral blood T cell antigens (25), and mouse sperm tail (26).

A 0.5% final concentration of NP40 was favoured (24, 25) and again a very simple extraction procedure worked adequately. Virus (24) or T cell (25) lysate was solubilised in 0.5% NP40 and then either centrifuged (retaining the soluble supernatant) in the case of T cell antigens, or stored at -70°C in the case of virus. The antigens were then used for attachment to the solid phase.

In the case of mouse sperm tail (26), NP40 was used to remove a glycoprotein (sperm maturation antigen) from mouse sperm tail. This glycoprotein was then assayed by ELISA without any apparent problems in the assay.

4. Problems with Detergents in ELISA Systems

A. Which detergent to use?

The first problem could well be the choise of detergent. Which detergent should be used? This aspect is well covered in various reviews (6, 7, 31, 33), and considers which detergents are likely to be most suited for a particular problem, e.g. DOC tends to dissociate interaction between proteins more effectively than nonionic detergents (7).

There are however no absolute rules for ELISA studies as to which detergent to use, but it is worth considering the following points.

1) SDS binds co-operatively to most membrane proteins. This tends to disrupt sub-unit structure and favours denaturation (34). Therefore, this detergent is good for solubilising and separating membrane components. However, it is worth pointing out that SDS may stabilise protein that is associated with the hydrophobic area of the phospholipid bilayer (31). Nevertheless, SDS must be considered a poor risk first choice detergent (31).

2) DOC does not usually denature proteins (6), but this must be viewed with some caution because it has been reported that DOC has inactivated several membrane enzymes (7). It does dissociate proteins (7) (but not always, 6, 34) i.e. it dissociates interactions between proteins but not as efficiently as does SDS. DOC tends to be used at pH values in excess of 7.8. However, DOC is not as 'safe' a detergent as are the nonionic detergents.

3) Cationic detergents are less likely to denature most proteins than SDS, but they have not been widely used in ELISA systems.

4) Little is known about Zwitterionic detergents in ELISA - they are in general effective at disaggregating hydrophobic proteins (18).

5) TX 100 and NP40 are nonionic and show even less interaction with proteins than anionic DOC (6). Nonionic detergents are very inefficient at breaking protein-protein interactions. TX 100 and NP40 (i.e. nonionic detergents with polyoxyethylene head groups) do not usually denature proteins. This is in their favour. Nonionic detergents are efficient at breaking lipid-lipid and lipid-protein interactions in membranes, but they are, as stated, inefficient at breaking protein-protein interactions (7). Nonionic detergents are regarded as being 'mild' in action (6, 7, 31, 33). The polyoxyethylene series (e.g. TX 100/NP40) do absorb ultraviolet light and interfere with the spectrophotometric measurement of protein. They are also subject to autoxidation, although this can be prevented. Nonionic detergents also permit ion exchange chromatography (31).

6) A further point to consider in the choice of detergent is that some organisms may be resistant to solubilisation with particular detergents. It is reported that nonionic detergents solubilise the cytoplasmic membranes of bacteria effectively, but that the outer membrane layers of Gram-negative bacteria are resistant to detergents. TX 100 removes little cell wall protein from _E. coli_, and only a portion of the lipopolysaccharide (6). This illustrates the point that, prior to selecting a detergent, consideration should be given to the type of structure (be it tissue, cell, virus, parasite, bacterium) which will be treated with that detergent.

B. What concentration of detergent to use?

This has already been dealt with in section 3. It should be emphasized here, however, that the final concentration of detergent in the "antigen" applied to the solid phase may inhibit this binding process (16).

It is possible that more efficient antigen adsorption to the solid phase from very dilute detergent extracts could be achieved by pretreatment of the wells of the standard 96 well microtitre tray by using the following procedure: 2.5% foetal calf serum (FCS), air drying, subsequently treating with a 2% glutaraldehyde solution, washing, and then applying antigen (in detergent) (25). Other methods may be useful (1, 35), but would require verification in detergent EIA systems.

Alternatively, detergent could conceivably be replaced either by another detergent or removed altogether (8, 17) by a variety of methods if it was found to interfere in the ELISA system under study. The complete removal of the detergent prior to antigen attachment, presupposes that the protein aggregation which occurs during detergent removal (6) will not have a deleterious effect on antigen binding or epitope expression on the solid phase.

Acknowledgements

The author would like to acknowledge the excellent secretarial assistance of Molly Puddephatt and Molly Reid of the NESCOT Typing Services.

References

1. Blake, C., Gould, B.J.: Use of enzymes in immunoassay techniques, a review. Analyst 109, 533-547 (1984).
2. Voller, A., Bartlett, A., Bidwell, D.E.: Enzyme immunoassays with special reference to ELISA techniques. J. clin. Path. 31, 507-520 (1978).
3. Weissman, D., Parker, D.J., Rothstein, T.L., Marshak-Rothstein, A.: Methods for the production of xenogeneic monoclonal antibodies in murine ascites. J. Immunol. 135 (2), 1001-1003 (1985).

4. Kurki, P.: Determination of anti-actin antibodies by a solid-phase immunoenzymatic assay and by indirect immunofluorescence technique. Clin. Immunol. Immunopathol. 11, 328-338 (1978).

5. Huikeshoven, H., Landheer, J.E., Denderen, A.C.v., Vlasman, M., Leiker, D.L., Das, P.K., Goldring, O.L., Pondman, K.W.: Demonstration of dapsone in urine and serum by ELISA inhibition. Lancet 1, (8058), 280-281 (1978).

6. Helenius, A., Simons, K.: Solubilisation of membranes by detergents. Biochem. biophys. Acta 415, 29-79 (1975).

7. Helenius, A., McCaslin, D.R., Fries, E., Tanford, C.: Properties of detergents. Meth. Enzymology 56, 734-749 (1979).

8. Furth, A.J.: Removing unbound detergent from hydrophobic proteins. Anal. Biochem. 109, 207-215 (1980).

9. Bio-Rad Laboratories Ltd., Watford, Hertfordshire, U.K.: 2-Dimensional polyacrylamide gel electrophoresis. Bulletin 1144, 1-11 (1985).

10. Jacobson, G.R., Saier, M.H. Jr.: Biological membranes: structure and assembly. In: "Biochemistry", Co-ordinating author, Zubay, G.L., Addison-Wesley Publishing Co. Inc., Massachusetts, California, London, Amsterdam, Ontario, Sydney, pp. 573-619 (1983).

11. Plumley, F.G., Schmidt, G.W.: Rocket and crossed immunoelectrophoresis of proteins solubilised with sodium dodecyl sulphate. Anal. Biochem. 134, 86-95 (1983).

12. Caterson, B., Baker, J.R., Levitt, D., Paslay, J.W.: Radioimmunoassay of the link proteins associated with bovine nasal cartilage proteoglycan. J. biol. Chem. 254, 9369-9372 (1979).

13. Thonar, E.J.-M., Kimura, J.H., Hascall, V.C., Poole, R.A.: Enzyme-linked immunosorbent assay analysis of the hyaluronate-binding region and the link protein of proteoglycan aggregate. J. biol. Chem. 257, 14173-14180 (1982).

14. Kammer, K.: Monoclonal antibodies to influenza A virus FM1 (H1 N1) proteins require individual conditions for optimal reactivity in binding assays. Immunology 48, 799-808 (1983).

15. Batteiger, B., Newhall, W.J., Jones, R.B.: The use of Tween 20 as a blocking agent in the immunological detection of proteins transferred to nitrocellulose membranes. J. Immunol. Meth. 55, 297-307 (1982).

16. Jeansson, S., Elwing, H., Nygren, H., Olofsson, S.: Evaluation of solubilised herpes simplex virus membrane antigens in diffusion in gel-enzyme-linked immunosorbent assay (DIG-ELISA). J. Virol. Meth. 4, 167-176 (1982).

17. Chin, J.C.: Comparison of different antigenic preparations for the detection of ovine serum antibodies against Brucella ovis by ELISA. Aust. vet. J. 60, (9), 261-264 (1983).

18. Perdew, G.H., Schaup, H.W., Selivonchick, D.P.: The use of a zwitterionic detergent in two-dimensional gel electrophoresis of trout liver microsomes. Anal. Biochem. 135, 453-455 (1983).

19. Hansen, E.S., Knudsen, J.: Limitations in the use of the enzyme-linked immunosorbent assay (ELISA) for identification and quantitation of thermogenin. J. Immunol. Meth. 77, 297-304 (1985).

20. Scheid, A., Choppin, P.W.: Isolation and purification of the envelope proteins of Newcastle disease virus. J. Virol. 11, (2), 263-271 (1973).

21. Nishikawa, K., Isomura, S., Suzuki, S., Watanabe, E., Hamaguchi, M., Yoshida, T., Nagai, Y.: Monoclonal antibodies to the HN glycoprotein of Newcastle disease virus. Biological characterisation and use for strain comparisons. Virology 130, 318-330 (1983).

22. Schmaljohn, A.L., Kokubun, K.M., Cole, G.A.: Protective monoclonal antibodies define maturational and pH-dependent antigenic changes in Sindbis virus E1 glycoprotein. Virology 130, 144-154 (1983).

23. Ziegler, T., Meurman, O.H., Arstila, P.P., Halonen, P.E.: Solid phase enzyme-immunoassay for the detection of herpes simplex virus antigens in clinical specimens. J. Virol. Meth. 7, 1-9 (1983).

24. Crouch, C.F., Raybould, T.J.G.: Comparison of different antigen preparations as substrates for use in passive haemagglutination and enzyme-linked immunosorbent assays for detection of antibody against bovine enteric coronavirus. J. Clin. Microbiol. 18, (1), 146-149 (1983).

25. Rabinovsky, E.D., Yang, T.-J.: Identification and quantitation of bovine T cell antigen with enzyme-linked immunosorbent assay (ELISA). J. Immunol. Meth. 56, 185-191 (1983).

26. Vernon, R.B., Hamilton, M.S., Eddy, E.M.: Effects of in vivo and in vitro fertilization environments on the expression of a surface antigen of the mouse sperm tail. Biol. Reprod. 32, 669-680 (1985).

27. Vaheri, A., Salonen, E.-M.: Evaluation of solid-phase enzyme-immunoassay procedure in immunity surveys and diagnosis of rubella. J. Med. Virol. 5, 171-181 (1980).

28. Tolskaya, E.A., Savinskaya, S.S., Balayan, M.S., Andzhaparidze, A.G., Kolesnikova, M.A.: Enzyme-immunoassay in the diagnosis of human hepatitis A: specific and non-specific reactions. Acta. virol. 27, 43-50 (1983).

29. Porstmann, B., Porstmann, T., Nugel, E.: Comparison of chromogens for the determination of horseradish peroxidase as a marker in enzyme immunoassay. J. Clin. Chem. Clin. Biochem. 19, 435-439 (1981).

30. Walden, P.W., Nagy, Z.A., Klein, J.: Induction of regulatory T-lymphocyte responses by liposomes carrying major histocompatibility complex molecules and foreign antigen. Nature 315, 327-329 (1985).

31. Tanford, C., Reynolds, J.A.: Characterisation of membrane proteins in detergent solutions. Biochim. biophys. Acta 457, 133-170 (1976).

32. Gurr, M.I., James, A.T.: Lipids as components of macromolecules. In: "Lipid Biochemistry", Chapman and Hall Ltd., Science Paperbacks, London, pp. 178-215 (1980).

33. Johnstone, A., Thorpe, R.: Isolation and fractionation of lymphocytes. In: "Immunochemistry in Practice", Blackwell Scientific Publications, Oxford, London, Edinburgh, Boston, Melbourne, pp. 77-101 (1982).

34. Freedman, R.B.: Membrane-bound enzymes. In: "New Comprehensive Biochemistry" vol. 1 (General Editors: Neuberger, A., Deenen, L.L.M. van) Membrane Structure, Eds. Finean, J.B., Michell, R.H., Elsevier North Holland Biomedical Press, Amsterdam, New York, Oxford, pp. 161-214 (1981).

35. Lehtonen, O.-P., Viljanen, M.K.: Antigen attachment in ELISA. J. Immunol. Meth. 34, 61-70 (1980).

DETECTION AND QUANTITATION OF SPERM ANTIBODIES BY A MODIFIED ELISA TECHNIQUE

H. Wolff, W.-B. Schill
Department of Dermatology, Andrology Unit, Ludwig-Maximilians-University of Munich, D-8000 Munich, Federal Republic of Germany

Introduction

When evaluating sera and genital secretions of infertile men and women, antisperm antibodies are more often found than among fertile controls (1-3). Using appropriate test systems, the sera, seminal plasma or cervical mucus of some patients cause agglutination (4, 5) or immobilization of spermatozoa (6), resulting in the inability of spermatozoa to penetrate cervical mucus (7). Other, more discrete alterations by sperm antibodies may impair the fertilizing capacity of spermatozoa (8).

However, most of the commonly used sperm antibody assays are of subjective evaluation, time-consuming to perform and require fresh donor semen of excellent quality as antigen (4-7).

To overcome these disadvantages, enzyme-linked immunosorbent assays (ELISAs) for the detection of sperm antibodies have been developed (9, 10).

This article describes a modified sperm antibody-ELISA using whole spermatozoa as antigens coated to positively charged polyvinylchloride microtiter plates.

In contrast to ELISAs used so far, fixation of spermatozoa to the microwells by glutaraldehyde can be avoided, thus protecting sperm antigens from potential alteration by this highly reactive substance (11).

Immunoassay Technology, Vol. 2

ELISA for the Detection of Sperm Antibodies

The principle of the indirect ELISA is shown in Fig. 1. To attain optimal test conditions, several variations have been performed.

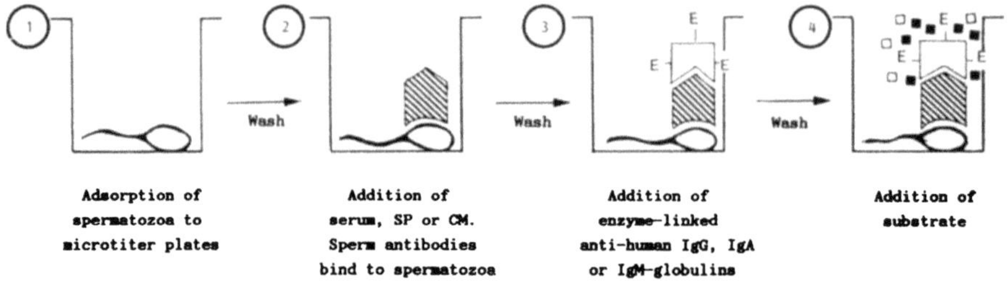

Fig. 1. Principle of the ELISA for the detection of antisperm antibodies. SP = Seminal Plasma; CM = Cervical Mucus.

Microtiter plates

Four types of commercial microtiter plates were compared for their ability to adsorb spermatozoa, to discriminate sperm antibody-positive from -negative samples, and to give reproducible results. The microtiter plates were made of polystyrene (Dynatech Lab., Denkendorf, W. Germany) or of polyvinylchloride (PVC), the latter being either positively or negatively charged, or ozone-activated (Titertek, Flow Lab., Meckenheim, W. Germany).

All three types of PVC microtiter plates were able to bind a high percentage of spermatozoa which was indicated by the absence of sperm cells in the supernatant fluid and washing buffer. Light-microscopy of the microwells showed single spermatozoa regularly disseminated over the well-bottoms.

On the other hand, polystyrene plates were not able to adsorb spermatozoa sufficiently unless glutaraldehyde was applied.

Optimal discrimination between sperm antibody-positive and -negative samples (i.e. biggest difference in absorbance values) was obtained with positively activated (charged) PVC microtiter plates (Titertek).

Moreover, intra- and inter-assay reproducibility was best with these plates. Thus, the variations reported in the following pages were all performed on positively charged PVC microtiter plates, if not otherwise indicated.

Spermatozoa as antigens

Human donor-semen obtained by masturbation and showing minimal requirements ($\geq 30 \times 10^6$ sp./ml, ≥ 50% total motility, ≥ 50% normal morphology, no leucocytes, no agglomerations) was subjected to the following procedures to obtain an appropriate antigen preparation.

After liquefaction, phosphate buffered saline (PBS) was added to the ejaculate until the volume was 10 ml. The suspension was centrifuged at 400 g for five minutes, the supernatant fluid decanted, the sperm pellet resuspended in 10 ml PBS and centrifuged again. This washing procedure was repeated twice. Finally, washed spermatozoa was resuspended in 2 ml PBS and stored frozen at $-20^{\circ}C$ until used.

To exclude sperm-antigen alteration by freezing, five washed sperm samples were subdivided and coated to the microtiter plates, both frozen and unfrozen. Evaluation of several sperm antibody-positive and -negative reference sera showed no differences between freshly coated and previously frozen spermatozoa as antigens.

In order to preclude interindividual sperm-antigen variations, sperm antibody-positive and -negative reference sera were applied to microtiter plates which were coated with the spermatozoa of five donors. As reported in the literature (3), there were no differences in the results obtained, using different spermatozoa.

Coating of microtiter plates with spermatozoa

Washed spermatozoa were adjusted to several sperm concentrations

(2.5, 5, 10, 20, 40, 80, 160, 320 x 10^3 sp./100 µl PBS) and 100 µl of the solutions were added to each well of the microtiter plates (Fig. 1-1) which were then centrifuged at 300 g for ten minutes.

In contrast to the PVC plates, which were ready for use immediately, the polystyrene plates had to be tapped out and 100 µl/well of 0.25% glutaraldehyde (in PBS) were added for ten minutes, followed by three washings with 200 µl PBS-Tween/well.

All sperm-coated plates were stored frozen at -20°C until used. No differences in results were seen when comparing untreated (PVC plates) and glutaraldehyde-fixed spermatozoa (polystyrene plates) as antigens.

However, in order to exclude any potential sperm-antigen alteration by glutaraldehyde and also to adopt a simpler way of coating, the positively activated PVC microtiter plates were chosen for further investigation.

As can be seen in Fig. 2, the adsorption capacity of these plates is so high that even amounts as small as 2.5 x 10^3 spermatozoa per well yield reasonable results. However, to be sure that there is enough antigen available, 100 x 10^3 should be coated into each microwell.

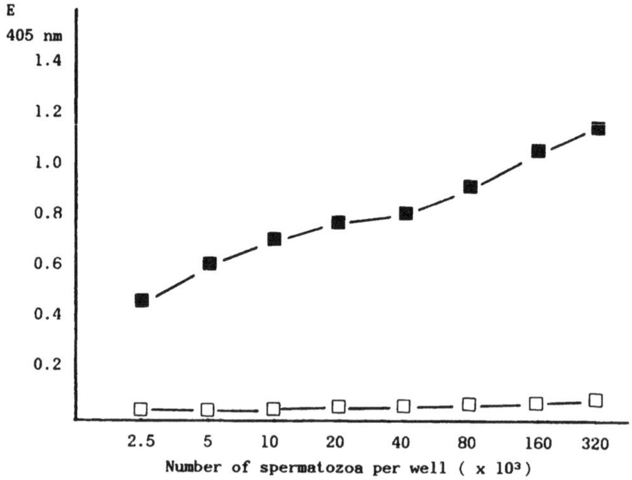

Fig. 2. Adsorption of spermatozoa to positively charged microtiter plates. Even as small amounts as 2.5 x 10^3 yield a clear distinction between a sperm antibody positive (■—■) and negative (□—□) serum.

Sera and seminal plasma samples

The infertile patients' seminal plasma was obtained by centrifugation of liquefied semen (16000 g, 3 min) about 60 to 90 minutes after masturbation.

Sera and seminal plasma were stored frozen at -20°C until used. To evaluate the influence of the complement system, 15 sperm antibody-positive and -negative sera were tested by the ELISA both activated (unheated) and inactivated (56°C, 30 min).

No differences in results were seen, indicating that the attachment of sperm antibodies to spermatozoa within the ELISA is not dependent on the complement system.

To exclude a major loss of antibody activity by multiple freezing and thawing, two positive sera and seminal plasma were subdivided and submitted to repeated freezing and thawing. No loss of antibody activity was detected in the ELISA up to six freezings. After ten freezings, sperm antibody activities of the test samples were still more than 90%.

Evaluation of cervical mucus

Evaluation of cervical mucus specimens for sperm antibodies by the ELISA technique requires some pretreatment in order to reduce the high viscosity. Liquefaction of mucus can be obtained either by treatment with the proteolytic enzyme bromelain (12) or by sonication (13).

However, as bromelain removes the Fc-part of immunoglobulins, the remaining Fab-parts bound to spermatozoal antigens cannot be detected by the heavy chain-specific antisera applied in the ELISA (14).

On the other hand, sonication of mucus does not affect immunoglobulins, thus being the method of choice for liquefaction of cervical mucus (13, 14).

It is recommended that an equal volume of PBS be added to the ovulatory cervical mucus sample (about 200-500 µl) and to sonicate in a Branson Sonifier for about 30 seconds until the mucus is liquefied. The pretreated cervical mucus can be stored frozen at -20°C until evaluated in the ELISA.

Performance of the ELISA

After thawing, the sperm-coated plates are washed with 200 µl PBS-Tween per well. Tween 20 (0.05%; Merck, Darmstadt, W. Germany) prevents proteins, i.e. antibodies, from non-specific binding to the walls of the microtiter plates; 100 µl of serum-, seminal plasma- or cervical mucus-dilutions (1/64, 1/32 or 1/4 in PBS-Tween) are added to the microwells (Fig. 1-2), all determinations being performed in duplicate. The plates are incubated for 90 minutes at 37°C in a moist chamber. After incubation, they are tapped out and washed three times with 200 µl PBS-Tween.

As second antibody, goat anti-human IgG, IgA or IgM conjugated to alkaline phosphatase is used (Sigma, Taufkirchen, W. Germany); 100 µl of a 1/1000 antibody-dilution (in PBS-Tween) are added to each well and incubated for a further 90 minutes (Fig. 1-3).

After three washings, 100 µl of the substrate p-nitrophenylphosphate (Sigma) dissolved in diethanolamine buffer (1 mg/ml) are added and incubated for 30 minutes at 37°C (Fig. 1-4).

The optical density of the yellow hydrolyzation-product is automatically measured at 405 nm by a multiscan plate reader (Titertek Multiskan MCC).

Controls

Different methodological controls were performed; wells without spermatozoa as antigens, wells incubated with PBS-Tween instead of serum, seminal plasma, enzyme conjugate or substrate.

Extinctions of these controls were below 0.015 after 30 minutes substrate-incubation, indicating a very low non-specific background activity.

Sperm antibody-positive and -negative reference samples

have to be used on each microtiter plate: WHO-reference sera (IgG, IgM), seminal plasma of a man with high titers of spermagglutinating antibodies (IgA), human γ-globulins (IgG) and sera, seminal plasma and cervical mucus of antibody-negative proven fertile men and women.

As described in the literature, we also observed a Fc-receptor-like non-specific binding of human IgG-globulins to the spermatozoal surface (15), giving a strong positive reaction in the assay. However, this was only the case with purified human γ-globulins (Cohn fraction IV, Sigma) diluted in 1 ml PBS but not with IgG from patients' sera. The fact that there was no correlation between increased IgG concentrations up to 2500 mg/dl serum and positive results in the ELISA is evidence against false positive results caused by raised IgG in the patients' serum. Thus, binding of IgG antibodies in patients' sera probably does not take place by the non-specific Fc-part but by the sperm-antigen specific Fab-parts.

Interpretation of the Results

A reaction is considered positive if the absorbance of a test sample is more than twice that of a negative control sample. The titer is the inverse of the highest sample-dilution still giving positive results (Fig. 3).

Another possible way of evaluating is by setting samples in relation to a positive standard sample of which the absorbance is regarded as 100%. The results of the test samples are expressed in per cent values of the positive standard (e.g. WHO-reference sera, human γ-globulins).

Sensitivity of the ELISA

Two positive WHO-reference sera were evaluated by different sperm antibody-assays (Table 1). The ELISA proved to be more

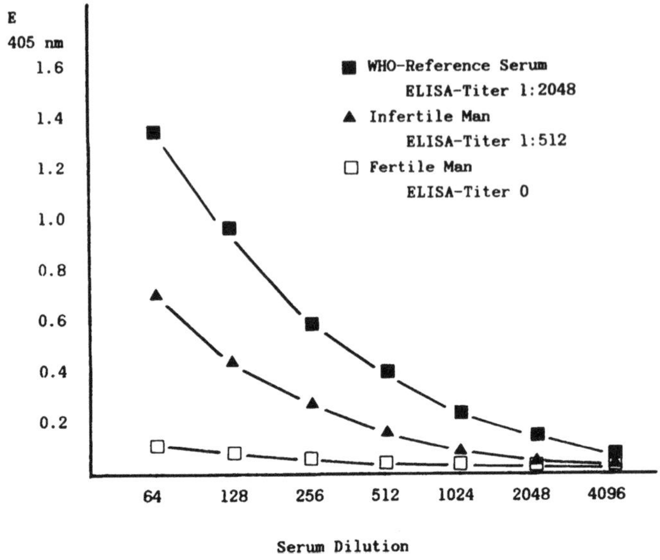

Fig. 3. Serum-titration curves of a sperm antibody-positive and negative man and a WHO-reference serum.

sensitive than the gelatine agglutination test (4), the sperm immobilization test (6) and the tray agglutination test (5).

Table 1. Evaluation of two WHO-reference sera by different sperm antibody assays. Highest titers of sperm antibodies were obtained by the ELISA technique.

WHO-Reference Serum	SIT (Isojima)	TAT (Friberg)	GAT (Kibrick)	ELISA
No. 108, Titer:	16	32	4	128
No. 116, Titer:	32	256	256	2048

SIT = Sperm Immobilization Test (6); TAT = Tray Agglutination Test (5); GAT = Gelatin Agglutination Test (4).

Reproducibility of the ELISA

Sperm antibody-positive and -negative sera and seminal plasma were each pipetted into 88 microwells of one microtiter plate. Intra-assay reproducibility (one day, one plate; 100% minus coefficient of variation) ranged from 88% to 94% and inter-assay reproducibility (different days and plates) from 72% to 76%.

Conclusions

Compared to commonly used sperm antibody-assays (4-6), the ELISA technique has a number of advantages. There is no need for fresh donor semen and even cryopreserved spermatozoa of average quality ejaculates may be used as antigens.

The ELISA proved to be more sensitive than the sperm immobilization test (6), the gelatine agglutination test (4), and the tray agglutination test (5; Table 1).

In addition, the reproducibility of the ELISA is very high, probably because many microtiter plates can be coated by one semen pool, thus allowing the performance of hundreds of determinations with identical antigens.

Furthermore, sperm-coated microtiter plates can be stored frozen for more than twelve months without any loss of quality.

Another advantage is the objective reading of the results by an automatic multiscan-spectrophotometer which contributes to a high degree of standardization. Due to the use of a 12-channel multipette (Titertek) for the washing procedures and to the speed of the optical reading (96 microwells within one minute), it is possible to screen more than one hundred test samples within one day.

In contrast to other sperm antibody ELISAs (9, 10), it is not necessary to fix spermatozoa to the microtiter plates with the help of glutaraldehyde when positively charged PVC microtiter plates are used. Thus, any sperm-antigen alteration by this highly reactive substance (11) can be excluded.

The ELISA is able to detect not only sperm-agglutination or immobilizing antibodies, but also antibodies possibly disturbing gamete-interaction, e.g. by reducing the ability of spermatozoa to attach and penetrate the zona pellucida of the egg (8).

Finally, by using heavy-chain specific IgG, IgA or IgM antisera, the immunoglobulin class of sperm antibodies can be determined.

Clinical Results obtained by the ELISA

Sera and seminal plasma of 178 andrological patients were evaluated by the modified ELISA. Sperm antibodies were found in 10% of the infertile men's sera and in 7% of their seminal plasma. It was striking that only one of 29 positive men had sperm antibodies both in serum and in seminal plasma, being evidence against a relevant transfer of circulating antibodies into genital secretions and attesting to the production of locally secreted sperm antibodies (17).

ELISA-reactive antibodies were found in 13% of the sera and in 15% of the cervical mucus samples of 40 infertile women. Similar to the results for infertile men, only one of 10 positive women had sperm antibodies both in serum and in cervical mucus (14).

Evaluating the sera of 223 dermatological patients without known fertility problems, sperm antibodies were found in nine cases (4%). Interestingly, six of the nine positive patients suffered from acute genital infections (syphilis, gonorrhea, soorbalanitis), supporting the hypothesis that local inflammation may induce the formation of antibodies against spermatozoa, possibly by promoting a pathological contact between the "alarmed" immune system and auto-immunogenic spermatozoa (18).

The highest incidence of circulating sperm antibodies was found among 42 homosexual men with 28% being positive in the

ELISA (19). The most probable explanation for this phenomenon is the frequently practised anal intercourse among these men, resulting in small lesions of the rectal mucosa by which spermatozoa gain access to the immune system of the passive partner (20).

Therefore, a humoral immune response against spermatozoal auto- and iso-immunogens is often found in homosexual men.

Finally, in an enlarged series of tests, we evaluated 89 homosexuals for a possible relationship between sperm antibodies and antibodies against the human T-cell lymphotropic virus, HTLV-III (21).

Thirtyone (35%) of the homosexuals were positive for sperm antibodies and 21 (24%) had HTLV-III antibodies, being evidence for an infection by the retrovirus which in some cases leads to the development of the acquired immune deficiency syndrome, AIDS (22).

Sixteen of 58 sperm antibody negative men (28%) were carriers of the HTLV-III virus, in contrast to only 5 of 31 sperm antibody positive men (16%).

Thus, we found no positive correlation between the occurrence of sperm antibodies and HTLV-III antibodies among homosexual men, although there are reports that sperm antibodies may cross-react with T-lymphocytes (24).

Acknowledgements

This work was supported by the Deutsche Forschungsgemeinschaft (Schi 86/ 7-5).

References

1. Shulman, S., Friedman, M.R.: Antibodies to spermatozoa. V. Antibody activity in human cervical mucus. Am. J. Obstet. Gynecol. 122, 101 (1975).

2. Beer, A.E., Neaves, W.B.: Antigenic status of semen from the viewpoints of the female and the male. Fertil. Steril. 29, 3-22 (1978).

3. Jones, W.R.: Immunological infertility: fact or fiction ? Fertil. Steril. 33, 577-586 (1980).

4. Kibrick, S., Belding, D.L., Merril, B.: Methods for the detection of antibodies against mammalian spermatozoa. Fertil. Steril. 3, 340-438 (1952).

5. Friberg, J.: A simple and sensitive micromethod for demonstration of sperm-agglutinating activity in serum from infertile men and women. Acta obstet. gynec. scand. (suppl.) 36, 21-29 (1974).

6. Isojima, S., Li, T.S., Ashitaka, Y.: Immunologic analysis of sperm-immobilizing factor found in sera of women with unexplained sterility. Am. J. Obstet. Gynecol. 101, 677-683 (1967).

7. Kremer, J., Jager, S.: The sperm-cervical mucus contact test: a preliminary report. Fertil. Steril. 27, 335-340 (1976).

8. Bronson, R., Cooper, G., Rosenfeld, D.: Sperm-specific isoantibodies and autoantibodies inhibit the binding of human sperm to the human zona pellucida. Fertil. Steril. 38, 724-729 (1982).

9. Ackerman, S.B., Wortham, J.W.E., Swanson, R.J.: An indirect enzyme-linked immunosorbent assay (ELISA) for the detection and quantitation of antisperm antibodies. Am. J. Reprod. Immunol. 1, 199-205 (1981).

10. Witkin, S.S.: Enzyme-linked immunosorbent assay (ELISA) for detection of antibodies to spermatozoa. Res. Reprod. 15, 1-2 (1983).

11. Bronson, G., Cooper, G., Rosenfeld, D.: Sperm antibodies: their role in infertility. Fertil. Steril. 42, 171-183 (1984).

12. Ingerslev, H.J., Poulsen, F.: Bromelain for liquefaction of cervical mucus in sperm antibody testing: its effect on sperm-agglutinating immunoglobulin G. Fertil. Steril. 35, 61-63 (1980).

13. Van Kooij, R.J.: Solubilization of cervical mucus with bromelain and mechanical means. Int. J. Fertil. 29, 81-84 (1984).

14. Wolff, H., Schill, W.-B.: Determination of sperm antibodies in human genital secretions by an ELISA technique. Arch. Gynecol. (1986) (in press).

15. Richards, J.M., Witkin, S.S.: Non-immune IgG binding to the surface of spermatozoa by disulphide rearrangement. Clin. Exp. Immunol. 58, 493-501 (1984).

16. Wolff, H., Schill, W.B.: A modified enzyme-linked immunosorbent assay (ELISA) for the detection of antisperm antibodies. Andrologia 17, 426-434 (1985).

17. Friberg, J.: Immunological studies on human sperm-agglutinating seminal fluid. Acta obstet. gynec. scand. (suppl.) 36, 65-71 (1974).

18. Bandhauer, K.: Immunreaktionen bei Fertilitätsstörungen des Mannes. Urol. Int. (Basel) 21, 247 (1966).

19. Wolff, H., Schill, W.-B.: Antisperm antibodies in infertile and homosexual men: relationship to serologic and clinical findings. Fertil. Steril. 44 (1985) (in press).

20. Mavligit, G.M., Talpaz, M., Hsia, F.T., Wong, W., Lichtiger, B., Mansell, P.W.A., Mumford, D.M.: Chronic immune stimulation by sperm alloantigens. Support for the hypothesis that spermatozoa induce immune dysregulation in homosexual males. JAMA 251, 237-241 (1984).

21. Gallo, R.C., Salahuddin, S.Z., Popovic, M., Shearer, G.M., Kaplan, M., Haynes, B.F., Palker, T.J., Redfield, R., Oleske, J., Safai, B., White, G., Foster, P., Markham, P.D.: Frequent detection and isolation of cytopathic retroviruses (HTLV-III) from patients with AIDS and at risk for AIDS. Science 224, 500-503 (1984).

22. Gottlieb, M.S., Schroff, R., Schanker, H.M., Weisman, J.D., Fan, P.T., Wolf, R.A., Saxon, A.: Pneumocystis carinii pneumonia and mucosal candidiasis in previously healthy homosexual men. New Engl. J. Med. 305, 1425-1431 (1981).

23. Mathur, S., Goust, J.M., Williamson, H.O., Fudenberg, H.H.: Antigenic cross-reactivity of sperm and T-lymphocytes. Fertil. Steril. 34, 469-476 (1980).

24. Wolff, H., Meurer, M., von der Helm, K., Schill, W.-B.: HTLV-III antibodies in homosexual men with and without sperm antibodies (in preparation).

IMMUNOASSAY REAGENTS : NOTES FROM THE EDITOR

From time to time, the Editor of this periodical will publish details of some commercially available immunoassay reagents. The first note gives the details of Alkaline Phosphatase and Peroxidase preparations manufactured by Biozyme Laboratories Limited, an independent British registered company which isolates and purifies enzymes for the diagnostic and pharmaceutical industries. The majority of their products are exported to the U.S.A. and they operate a rigorous Research and Development programme aimed at significant improvements in their range of enzymes. Recently, Alkaline Phosphatase from calf intestine and Peroxidase from horse radish root have been upgraded for use in enzyme-linked immunoassay systems. The various grades currently available, together with the specifications are as follows:

(A) Alkaline Phosphatase (Calf Intestine)

Product Code	Glycine Buffer pH 9.6, 25°C Biozyme Assay	Glycine Buffer pH 10.4 37°C	Diethanolamine Buffer pH 9.8 37°C
ALPI-10G	>1300 U/mg protein	>1700 U/mg protein	>4200 U/mg protein
ALPI-9G	>1100 U/mg protein	>1400 U/mg protein	>3500 U/mg protein
ALPI-8G	> 800 U/mg protein	>1000 U/mg protein	>2500 U/mg protein

The above materials consist of a highly purified salt-free stabilised solution in 50% glycerol/0.5 mM $MgCl_2$/0.05 mM $ZnCl_2$. The solutions do not require dialysis prior to conjugation with antibodies. The ALPI-10G has a very high specific activity and is extremely stable in 50% glycerol. Alkaline Phosphatase can

Immunoassay Technology, Vol. 2

also be supplied in the freeze dried form at different levels of activity and in suspension in ammonium sulphate solution. Details are available upon request.

(B) Peroxidase (Horse Radish Root)

Product Code	Guaiacol Assay 25°C	I.U.B. 25°C	Purpurogallin 20°C	RZ
HRP 4	280 U/mg material	#3500 U/mg material	280 U/mg material	>3.0

This grade is supplied as a chromatographically salt-free freeze dried powder. Peroxidase can also be supplied at other levels of activity. Details are available upon request.

Available from: Biozyme Laboratories Limited, Unit 6, Gilchrist-Thomas Estate, Blaenavon, Gwent NP4 9RL, South Wales, Great Britain. Telephone: Blaenavon 790678 (STD Code 0495), Telegrams: Biozyme, Blaenavon, Telex: 497731.

CONTRIBUTORS

Numbers in parentheses indicate the page on which the authors' articles begin

W. B. Ershler, Vermont Regional Cancer Center and Department of Medicine, University of Vermont, Burlington, Vermont, U.S.A. (105).

O. L. Goldring, Department of Biological Sciences, North East Surrey College of Technology, Reigate Road, Ewell, Epsom, Surrey, KT17 3DS, U.K. (189).

D. Haidukewych, Epilepsy Center of Michigan, 3800 Woodward Avenue, Detroit, Michigan 48201, U.S.A. (71).

D. Monroe, Department of Medicine, Infectious Diseases and Connective Tissues Sections, University of Tennessee, Center for the Health Sciences, Memphis, Tennessee 38163, U.S.A. (57), (167).

Ann L. Moore, Vermont Regional Cancer Center and Department of Medicine, University of Vermont, Burlington, Vermont, U.S.A. (105).

W.-B. Schill, Department of Dermatology, Andrology Unit, Ludwig-Maximilians-University of Munich, D-8000 Munich, Federal Republic of Germany (215).

R. Smeenk, Department of Autoimmune Diseases, Central Laboratory of the Netherlands Red Cross, Blood Transfusion Service, and Laboratory for Experimental and Clinical Immunology, University of Amsterdam, Amsterdam, The Netherlands (121), (145).

T. J. Wilkin, Department of Medicine II, University of Southampton, Southampton General Hospital, Southampton SO9 4XY, U.K. (17).

D. G. Williams, Department of Clinical Biochemistry, District General Hospital, Kayll Road, Sunderland, U.K. (1).

H. Wolff, Department of Dermatology, Andrology Unit, Ludwig-Maximilians-University of Munich, D-8000 Munich, Federal Republic of Germany (215).

NOTES ON CONTRIBUTORS

WILLIAM B. ERSHLER is a Clinical Research Oncologist/Immunologist at the Vermont Regional Cancer Center, University of Vermont. He is currently investigating aspects of aging, immunedeficiency and neoplasia. A native of Syracuse, New York, he completed his medical education at the State University of New York, Upstate Medical Center, and his research training at the University of Wisconsin, Madison.

DR. OWEN LESLIE GOLDRING is Research Director of the Department of Biological Sciences, North East Surrey College of Technology (NESCOT), Ewell, Surrey, U.K. This department, which he joined in September 1980, is one of the largest in the U.K. in the public sector of Further and Higher education, and has a wide research base. Dr. Goldring has three major areas of research interest: mechanism of antigen presentation and use of idiotypes/anti-idiotypes in experimental leishmaniasis; development of immunoassay systems for Leishmania, Eimeria, Aspergillus, and human haemoglobin sub-groups (a particular interest is the characterisation and use of detergent extracted antigens); scale-up of monoclonal antibody producing hybrids in tissue culture. Dr. Goldring completed a PhD (1972-1975) at the National Institute for Medical Research, Mill Hill, London, where he subsequently worked as a scientist between 1975-1976; he was Lecturer in Immunology at the International Immunology Training and Research Centre (ITR) in Amsterdam

(1976-1977) helping establish an MSC Immunology, and was ODA Lecturer in the Liverpool School of Tropical Medicine from 1977 to 1980; in 1979, he acted as Consultant to the Indian Council for Medical Research, on the use of ELISA for the immunodiagnosis of visceral leishmaniasis.

DR. DAN (BOHDAN) HAIDUKEWYCH is the Clinical Chemistry and Pharmacology Laboratory Director at the Epilepsy Center of Michigan, Detroit, and a Consultant to the local hospital laboratories. His research interests include clinical pharmacology of drug interactions, effect of anticonvulsants on serum enzyme status, and development of methodologies for therapeutic drug monitoring. In the past five years he has published extensively in journals such as Clinical Chemistry, Therapeutic Drug Monitoring, Neurology, and has chapters in Advances in Epileptology, XIIth and XVth Epilepsy International Symposiums. Dr. Haidukewych received the B.A. degree (1960) with distinction, from the University of Michigan, and the M.S. (1965) and Ph.D. (1970) degrees in Chemistry from the University of Detroit. He did two years of post-doctoral research studies at the University of Toronto and Wayne State University. He is a member of the American Chemical Society, The American Association for Clinical Chemistry, the American Epilepsy Society and the Ukrainian-American Engineers Society of North America.

DAN MONROE is a Research Associate investigating infectious diseases and connective tissue studies at the University of Tennessee Center for the Health Sciences (UTCHS) Memphis, Tennessee, U.S.A. He attended UTCHS medical school after receiving his BS degree in biology from Christian Brothers

College (Memphis, Tennessee). He is a scientific free lance writer, book reviewer, and for the past 19 years a member of the streptococcal infectious disease research team. His major interests include: protein chemistry, immunoassays, streptococcal pathogenesis, rheumatic fever, autoimmune diseases, and synthetic vaccine production. His present work has expanded to include collagen studies at the Veterans Administration Medical Center, Connective Tissue Section, Memphis, Tennessee.

ANN L. MOORE is a Research Scientist at the Vermont Regional Cancer Center, University of Vermont. Her research interests there have focused on *in vitro* assessment of immune function. In particular, she has been investigating techniques of measuring antibody synthesis. Born and raised in Berkeley, California, she completed her formal education at the University of California at Davis.

WOLF-BERNHARD SCHILL was born on November 10, 1939 in Bernburg/Saale and is Professor of Dermatology, Venerology and Andrology. He graduated from the University of Tuebingen in 1965, and was a Research Fellow and Research Associate in Biochemistry and Pharmacology at the Max-Planck-Institute for Experimental Medicine in Goettingen,

West Germany, from 1967 to 1969. From 1970 to 1971, Professor Schill was Ford Foundation Research Fellow in Reproductive Biology at the Department of Obstetrics and Gynecology of the University of Chicago, U.S.A. He then joined the Department of Dermatology at the Ludwig-Maximilians-University of Munich, completed his residency, and, in 1975, became a licensed dermatologist. In 1976, he was appointed as Lecturer in Dermatology, Venerology and Andrology, and became Assistant Professor. Since 1980 he has been Associate Professor at the Department of Dermatology, University of Munich. Professor Schill is a member of several national and international scientific societies. His main fields of research are clinical andrology, reproductive biology, biochemistry of semen, cryobiology, proteinases and proteinase inhibitors, immuno-reproduction and antienzymatic contraception.

From 1971 to 1978, RUUD SMEENK studied biochemistry with Professor Arthur Rörsch and immunology with Professor Jan van Rood, both at Leiden University. Towards the end of 1978, he went to Amsterdam, to work in the laboratory of Professor Bert Feltkamp and Lucien Aarden (Central Laboratory of the Netherlands Red Cross Blood Transfusion Service, Department of Autoimmune Diseases). Here, he was involved in the technology of anti-DNA detection, and studied the role of anti-DNA avidity in various assay systems. He published several papers on this subject and wrote a thesis entitled "Low Avidity antibodies to dsDNA", published by Rodopi, Amsterdam, for which he obtained his Ph.D. in 1982. Since then, he has published several papers with Professor Carel van Oss from the University of New York at Buffalo, on the biophysics of DNA/anti-DNA interaction. Dr. Smeenk studied complement fixation and activation of anti-DNA with Rafael Herrera Esparza from the Universidad

Autonoma de Zacatecas, Mexico. Clinical relevance of anti-DNA detection by several assays and the clinical relevance of anti-DNA avidity was, and still is, studied (and published) together with Tom Swaak, who is the head of a rheumatology clinic in Rotterdam. Last year, the author took part in the organisation of the IVth Bertine Koperberg Conference on Antinuclear Antibodies and Antigens and was co-editor of the Proceedings of this Meeting (Scand. J. Rheumatol. suppl. 56). At present, the author still works in Amsterdam, both on anti-DNA and on Antinuclear Antibodies, using monoclonal antibodies to study nuclear antigens, anti-DNA cross-reactivity and behaviour of (monoclonal) antibodies in assay systems.

TERENCE WILKIN graduated from the University of St. Andrew's Medical School in 1969 and trained in endocrinology under the late Professor James Crooks. He received his MD on thyroid autoimmunity in 1978 and spent the years 1978-1981 with Professors Claude Jaffiol and Jacques Mirouze at the University of Montpellier, France, first as Wellcome Travelling Fellow and subsequently as Associate Professor. He returned to the U.K. in 1981 and is currently Wellcome Senior Lecturer in Endocrinology in the Department of Medicine II at the University of Southampton. Dr. Wilkin has published some 45 papers, mainly on endocrine autoimmunity, and currently leads a research group investigating insulin autoimmunity.

D.G. WILLIAMS was born in Porth, Glamorgan, in 1948 and educated at Tonypandy Grammar School and Hawardian High School for Boys, Cardiff, Wales, U.K. He graduated from the University College of Wales, Aberystwyth, in 1970, with an honours degree in

Biochemistry. He proceeded to the Master of Science course in Clinical Biochemistry at the University of Birmingham, under Professor T.P. Whitehead. Following graduation in 1972, he worked in London (Northwick Park Hospital and Clinical Research Centre) and Newcastle-upon-Tyne, before obtaining the Top Grade Biochemist post in Sunderland in 1984.

HANS WOLFF was born on April 25, 1959 in Weidenbach/Romania and graduated in 1986 from the Ludwig-Maximilians-University of Munich. In 1983 he joined the Andrology Unit of the Department of Dermatology, University of Munich, to work on his thesis in the field of immunoreproduction. Dr. Wolff established an enzyme-linked immunosorbent assay for the detection of antisperm antibodies and, by this technique, assessed the sera and genital secretions of infertile men and women as well as the sera of dermatological patients and homosexual men. In addition, investigations were performed on the significance of sperm antibodies among HTLV-III-positive and -negative homosexual men.

SUBJECT INDEX

Contents of Volume 1.

Instructions to Authors for Preparing Manuscripts

Most recently published literature on developments in immunoassay technology (raising of antisera, preparation and labelling of reagents, new immunoassay methods, immunobiology, immunoelectrophoresis, immunofluorescent analysis, immunoelectron microscopy, immunocytology, immunogenetics, immunohistochemistry, immunopharmacology, instruments and automation) of biological, environmental, industrial, medical and pharmaceutical importance, should be reviewed in clear concise English free from jargon, justifying the title of this BIANNUAL SERIES. Spelling should follow the Oxford Dictionary of Current English. The reviewer's own findings, if any, published or in press, could also be mentioned BRIEFLY.

Should an investigator wish to publicise his own work, he should submit his written manuscript as a scientific paper.

Contributors must submit their typewritten papers in duplicate, not normally exceeding 50 A4 size (21.0 cm x 23.7 cm) pages including diagrams, figures, tables and references, using sub-headings for the topics discussed, and typed with double spacing on one side only of the paper. The typescript must be letter perfect as no galley proofs will be sent to the contributors, since the edited manuscripts will be retyped on reproduction paper for photo-printing.

All diagrams, plates, tables and other illustrations, including the caption/legend, which should make their meaning clear without reference to the text, must be photographed with well contrasted print, on glossy paper, and submitted in duplicate, each labelled with soft pencil on the back with the contributor's name and figure number and "top" edge identified.

The contributor should also indicate the exact size and place where the photographs of the diagrams, plates, tables or any other illustration should be inserted in the manuscript.

Text references should be numbered serially, as cited, without mentioning the names of the authors. The reference list at the end of the paper should be typed numerically, according to the text reference numbers (not alphabetically) including the names of ALL the authors, starting with the surname and followed by all the initials, full titles of articles, journal titles, volume number, first and last page numbers, and year of publication, for example:

Exley, D., Woodhams, B.: The specificity of antisera raised by oestradiol-17β-3-hemisuccinate-BSA. Steroids 27, 813-820 (1976).

Watson, R.A.A., Landon, J., Shaw, E.J., Smith, D.S.: Polarisation fluoroimmunoassay of Gentamicin. Clin. chim. Acta. 73, 51-55 (1976).

Reference to books should include chapter title, name of book, name of editors, publishers, place of publication, date of publication, edition, first and last page numbers, for example:

Conn, P.M., Bates, M.D., Rogers, D.C., Seay, S.G., Smith, W.A.: GnRH-Receptor-effector-response coupling in the pituitary gonadotrope: A Ca^{2+} mediated system. In: "The Role of Drugs and Electrolytes in Hormonogenesis", Eds. Fotherby, K., Pal, S.B., Walter de Gruyter & Co., Berlin/New-York, pp. 85-103 (1984).

The author should also underline with RED INK in the text not more than 20 keywords, meant for the subject index.

A recent photograph, STRICTLY 6 cm x 5 cm, in black and white, and brief details of the contributor's career, not exceeding 200 words, should also be enclosed with the manuscript.

Contributors should follow these instructions carefully while preparing their manuscripts to facilitate speedy publication of the book.

One contributor only for each article (first author or the senior author in case of a multiauthored article, whose name should be clearly indicated to avoid confusion) will receive a free copy of the book, 20 reprints and DM 10.- per printed page as the cost of preparing the manuscript. Further copies of the book will be sold to all contributors by the Publisher at 30% discount off the selling price.

Manuscripts, in duplicate, should be sent to:

Dr. S.B. Pal,
Universität Ulm, Department für Innere Medizin,
Steinhövelstrasse 9,
D-7900 Ulm (Donau) - Federal Republic of Germany.
(Telephone: Ulm (0731) 1791, extn. 2411).

Bei Fragen zur Produktsicherheit wenden Sie sich bitte an:
If you have any questions regarding product safety,
please contact:

Walter de Gruyter GmbH
Genthiner Straße 13
10785 Berlin
productsafety@degruyterbrill.com